T0302106

PERFUSION CELL CULTURE PROCESSES FOR BIOPHARMACEUTICALS

Master the design and operation of perfusion cell cultures with this authoritative reference. Discover the current state of the art in the design and operation of continuous bioreactors, with emphasis on mammalian cell cultures for producing therapeutic proteins. Topics include the current market for recombinant therapeutic proteins, current industry challenges, and the potential contribution of continuous manufacturing. The volume provides coverage of every step of process development and reactor operation, including small-scale screening to lab-scale and scale-up to manufacturing scale. Illustrated through real-life case studies, this is a perfect resource for groups active in the cell culture field, as well as graduate students in areas such as chemical engineering, biotechnology, chemistry, and biology, and to those in the pharmaceutical industry, particularly biopharma, biotechnology, and food or agro industry.

Moritz Wolf is a postdoctoral fellow at ETH Zürich in the department of Chemistry and Applied Biosciences.

Jean-Marc Bielser is an associate manager in the Biopharma Technology and Innovation group at Merck Serono SA (Switzerland). He obtained his master's degree in chemical engineering and biotechnology from EPFL, and the degree of Doctor of Science from ETH Zürich.

Massimo Morbidelli is Professor Emeritus in the Department of Chemistry and Applied Biosciences at ETH Zürich and Professor at the Department of Chemistry, Materials and Chemical Engineering at the Politecnico di Milano. Member of the Italian Academy of Sciences (Accademia dei Lincei), he received the Excellence in Process Development Research Award in 2017 from the American Institute of Chemical Engineers, the 2018 Separations Science and Technology Award from the American Chemical Society, and the 2019 Award in Integrated Continuous Biomanufacturing. He is the coauthor of *Continuous Biopharmaceutical Processes* (Cambridge University Press, 2018) and *Parametric Sensitivity in Chemical Systems* (Cambridge University Press, 2005).

Perfusion Cell Culture Processes for Biopharmaceuticals

Process Development, Design, and Scale-Up

Moritz Wolf

ETH Zürich

Jean-Marc Bielser

Merck Serono SA
ETH Zürich

Massimo Morbidelli

Politecnico di Milano
ETH Zürich

CAMBRIDGE
UNIVERSITY PRESS

University Printing House, Cambridge CB2 8BS, United Kingdom

One Liberty Plaza, 20th Floor, New York, NY 10006, USA

477 Williamstown Road, Port Melbourne, VIC 3207, Australia

314-321, 3rd Floor, Plot 3, Splendor Forum, Jasola District Centre, New Delhi - 110025, India

79 Anson Road, #06-04/06, Singapore 079906

Cambridge University Press is part of the University of Cambridge.

It furthers the University's mission by disseminating knowledge in the pursuit of
education, learning and research at the highest international levels of excellence.

www.cambridge.org
Information on this title: www.cambridge.org/9781108480031
DOI: 10.1017/9781108847209

First published 2020

A catalogue record for this publication is available from the British Library

Library of Congress Cataloging in Publication data
Names: Wolf, Moritz, author. | Bielser, Jean-Marc, author. | Morbidelli,
    Massimo, author.
Title: Perfusion cell culture processes for biopharmaceuticals : process
    development, design, and scale-up / Moritz Wolf, Jean-Marc Bielser,
    Massimo Morbidelli.
Other titles: Cambridge series in chemical engineering.
Description: New York, NY : Cambridge University Press, 2020. | Series:
    Cambridge series in chemical engineering | Includes bibliographical
    references and index.
Identifiers: LCCN 2020000228 (print) | LCCN 2020000229 (ebook) | ISBN
    9781108480031 (hardback) | ISBN 9781108847209 (epub)
Subjects: MESH: Batch Cell Culture Techniques | Perfusion–methods |
    Bioreactors | Biological Products–chemistry | Models, Chemical
Classification: LCC RM301.25 (print) | LCC RM301.25 (ebook) | NLM QS 525 |
    DDC 615.1/9–dc23
LC record available at https://lccn.loc.gov/2020000228
LC ebook record available at https://lccn.loc.gov/2020000229

ISBN 978-1-108-48003-1 Hardback

Contents

Contents

Abbreviations

ATF	alternating tangential flow
ATP	adenosine triphosphate
BPOG	Biophorum Operations Group
BR	benchtop bioreactor
CCC	critical coagulation concentration
CD	chemically defined
CFB	concentrated fed-batch
CFD	computational fluid dynamics
CGI	chemical growth inhibitor
CMP-Neu5Ac	cytidine diphosphate N-acethyneuraminic acid
COG	cost of goods
CPP	critical process parameter
CQA	critical quality attributes
$CSPR_{min}$	minimum cell-specific perfusion rate
CSPR	cell-specific perfusion rate
CSTR	continuous stirred tank reactor
DNA	deoxyribonucleic acid
DO	dissolved oxygen
DWP	deepwell plate
EGI	environmental growth inhibitor
ER	endoplasmic reticulum
ESS	explained variance
Fc	fragment crystallisable region
FDA	Food and Drug Administration
FucT	$\alpha - 1, 6$ fucosyltransferase
G0	no galactose molecule attached
G1	one galactose molecule attached
G2	two galactose molecules attached
GalT	$\beta - 1, 4$ Galactosyltransferase
GDP-Fuc	guanosine diphosphate fucose
GnTI	$\alpha - 1, 3$ N-acetylglucosaminyl transferase I
GnTII	$\alpha - 1, 6$ N-acetylglucosaminyl transferase II
HMW	high molecular weight
HS	high seeding fed-batch
IgG	immunoglobulin G
IPC	in-process control

List of Abbreviations

LCA	life-cycle assessment
LMW	low molecular weight
LS	low seeding fed-batch
LV	latent variable
mAb	monoclonal antibody
MALDI-TOF	matrix assisted laser desorption ionisation – time of flight
ManI	$\alpha - 1, 2$ mannosidase I
ManII	$\alpha - 1, 6$ mannosidase II
MAN	Mannose
MCSGP	multicolumn countercurrent solvent gradient purification
MIR	mid-infrared
MS	mass spectroscopy
msBR	micro-scale bioreactor
MVDA	multivariate data analysis
NIH	National Institure of Health
NIPALS	non-linear iterative partial least square
NIR	near infrared
NPV	net present value
NS	nucleotide activated sugar
NTP	nucleotide triphosphate
OS	oligosaccharide
OTR	oxygen transfer rate
PAT	process analytical technology
PCA	principle component analysis
PDE	partial differential equation
PF	perfusion
PFR	plug flow reactor
PID	proportional integral derivative
PLS	partial least square
PTM	post-translational modification
PMMA	poly(methyl methacrylate)
QbD	quality by design
relRMSEP	relative root mean square error in prediction
REMSECV	root mean square error in cross-validation
RMSEP	root mean square error in prediction
ROS	radical oxydative species
RT	Rushton turbine
RTD	residence time distribution
RV	reactor volume
SCADA	supervisory control and data acquisition
SialT	$\alpha - 1, 6$ sialyltransferase
ST	shake tube
STD	standard deviation
SVD	single value decomposition
TCA	tricarboxylic acid cycle
TFF	tangential flow filtration

List of Abbreviations

TMP	transmembrane pressure
TSS	total variance
U	uridine
UDP-Gal	uridine diphosphate galactose
UDP-GlcNAc	uridine diphosphate N-acetylglucosamine
UFDF	ultrafiltration diafilatration
UV	ultraviolet
VCD	viable cell density
VCD$_{max}$	maximum viable cell density
VVD	vessel volume per day

Symbols

$(\bar{\epsilon}_T)_g$	Average total energy dissipation rate, $W \times m^{-3}$
$(\bar{\epsilon}_T)_{Ig}$	Average gasing energy dissipation rate, $W \times m^{-3}$
$(\bar{\epsilon}_T)_S$	Average stirring energy dissipation rate, $W \times m^{-3}$
$\bar{\epsilon}_T$	Specific energy dissipation rate, $W \times m^{-3}$
ΔC	Distance between two impellers, m
ΔC_{Gas}	Gas driving force, $mol \times L^{-1}$
Δ	Diagonal matrix of the non-zero singular values, –
$\hat{y}_{test,i}$	Model estimation of the y-value of the ith observation, –
λ	Eigenvalue, –
μ	Cell growth rate, d^{-1}
μ_d^{max}	Maximum cell death rate, d^{-1}
μ_d	Cell death rate, d^{-1}
μ_L	Dynamic viscosity, $kg \times m^{-1} \times s^{-1}$
μ_l	Cell lysis rate, d^{-1}
μ_{max}	Maximum cell growth rate, s^{-1}
ω_j	Width of the concentration profile of E_j, –
ω_k	Width of the concentration profile of TP_k, –
$\omega_{q,mAb}$	Width of the specific productivity as a function of pH, –
ρ_L	Liquid density, $kg \times m^{-3}$
σ_L	Liquid surface tension, $N \times m^{-1}$
τ	Average residence time, s
τ_{max}	Maximum tolerable stress, $N \times m^2$
τ_{Sep}	Average residence time in the cell retention device, s
θ_m	Characteristic mixing time, s
A	Bioreactor cross section, m
A	Number of principal components, –
a	Gas–liquid interfacial area per unit dispersion volume, m^{-1}
B	Bleed rate, $L \times d^{-1}$
B	PLS regression coefficient, –
C	Weights of matrix Y, –
C_{Gas}^*	Saturated gas concentration, $mol \times L^{-1}$
$C_{O_2}^*$	Oxygen concentration at saturation in liquid phase, $mg/L, ppm$
c_i^0	Initial molar concentration of species i, $mol \times L^{-1}$
C_{Gas}	Gas concentration in the reactor, $mol \times L^{-1}$
$C_{Harvest}$	Protein concentration in the harvest stream, $g \times L^{-1}$
c_i	Molar concentration of species i, $mol \times L^{-1}$
C_{O_2}	Oxygen concentration in liquid phase, $mg/L, ppm$

List of Symbols

C_P	Protein concentration, $g_{Protein} \times d^{-1}$
$C_{Reactor}$	Protein concentration in the reactor, $g \times L^{-1}$
D	Impeller diameter, m
d	Cell diameter, μm
D_i	Golgi diameter, μm
E	Residual matrix of the X space, $-$
$E(t)$	Residence time distribution, s
E_j^{max}	Peak concentration of glycosyltransferase j, $mol \times L^{-1}$
E_j	Glycosyltransferase j, $-$
F	Residual matrix of the Y space, $-$
f	Frequency, s^{-1}
F_i^{in}	Molar flowrate of species i entering the bioreactor, $mol \times d^{-1}$
f_{inh}	Term indicating inhabation, $-$
F_i	Molar flowrate of species i leaving the bioreactor, $mol \times d^{-1}$
f_{lim}	Term indicating nutrient limitation, $-$
$F_{T,k}$	Flowrate of sugar precursors into the Golgi, $mol \times s^{-1}$
G	Residual matrix of the regression model, $-$
g	Acceleration of gravity, $m^2 \times s^{-1}$
G_i	Rate of production of species i, $mol \times d^{-1}$
H	Bioreactor height, m
H_L	Filling height of the cell culture broth, m
I_n	Concentration of growth inhibitor n, $mol \times L^{-1}$
k	Reaction rate constant, $mol \times L^{-1} \times s^{-1}$
$K_{UDP-Gal,Gal}^{Gal}$	Equilibrium constant of the UDP-Gal equilibrium, $mol \times L^{-1}$
$k_{f,j}^{max}$	Maximum turnover rate of a specific reaction, s^{-1}
$K_{NS,k}^{MS}$	Equilibrium constant describing the equilibrium between monosaccharide in the medium and in the cytosol, $mol \times L^{-1}$
$K_{\mu,AMM}$	Ammonia growth inhibition constant, $mol \times L^{-1}$
$K_{d,AMM}$	Ammonia death inducing constant, $mol \times L^{-1}$
$K_{d,i}$	Dissociation constant of the specific donor-enzyme complex, $mol \times L^{-1}$
$K_{d,Mn^{2+}}$	Dissociation constant of the specific manganese-enzyme complex, $mol \times L^{-1}$
$K_{d,Nk}$	Dissociation constant of the nucleotide-enzyme complex, $mol \times L^{-1}$
$k_{f,j}$	Turnover rate constant, s^{-1}
k_L	Gas-liquid mass transfer coefficient, $m \times s^{-1}$
$k_L a$	Volumetric mass transfer coefficient, s^{-1}
K_n	Monod constant, $kg \times L^{-1}$
$k_{T,k}$	Transport turnover rate, s^{-1}
M	Measured torque on the impeller shaft, $n \times m$
M	Number of variables constituting the data matrix X, $-$
m_{AMM}	Ammonia-maintenance-related coefficient, $mol \times d^{-1}$
$m_{NS,k}$	Nucleotide-sugar-maintenance-related coefficient, $mol \times d^{-1}$
$m_{UDP-Gal}$	UDP-Gal maintenance coefficient, $mol \times d^{-1}$
MC	Medium consumption, $L_{Medium} \times g_{Protein}$
MS_k	Concentration of monosaccharide in the medium, mol
N	Agitation speed, s^{-1}

List of Symbols

N	Number of observations of the data matrix X, –
N_A^{Golgi}	Ammonia-associated Golgi constant, $mol \times L^{-1}$
N_C	Number of species, –
N_i	Number of moles of species i, mol
N_k	Nucleotide k, –
N_R	Number of reactions, –
NR	Number of enzymatic reactions, –
NS_k	Nucleotide sugar k, –
OS_i	Oligosaccharide i, –
P	Perfusion rate, d^{-1}
P	Power input, W
P_0	Power number, –
p_a	Loading vector corresponding to the ath principal component, –
P_g	Gaseous power dissipation, W
$p_{m,a}$	Element of loading matrix P corresponding to ath principal component and mth variable, –
pK_A^{Golgi}	Apparent pK_A value of the Golgi, –
PR	Volumetric productivity, $g_{Protein} \times L_{Reactor} \times d^{-1}$
Q	Volumetric flowrate, $L \times d^{-1}$
Q^2	Relative variance explained in cross validation, –
Q_B	Bleed volumetric flowrate, $L \times d^{-1}$
Q_g	Volumetric gas flowrate, $L \times min^{-1}$
Q_H	Harvest volumetric flowrate, $L \times d^{-1}$
Q_{in}	Volumetric flowrate of nutrient feed addition, $L \times d^{-1}$
q_i	Specific production rate of species i, $mol \times L^{-1} \times d^{-1}$
q_{mAb}	Cell-specific productivity of monoclonal antibody, $g_{mAb} \times cell^{-1} \times d^{-1}$
Q_{out}	Volumetric flowrate of nutrient removal, $L \times d^{-1}$
Q_P	Perfusion volumetric flowrate of nutrient addition, $L \times d^{-1}$
q_p	Cell-specific productivity, $g_{Protein} \times cell^{-1} \times d^{-1}$
R	Overal rate of reaction, $mol \times L^{-1} \times d^{-1}$
r_i	Rate of production of species i, $mol \times L^{-1} \times d^{-1}$
R_j	Rate of the reaction j, $mol \times L^{-1} \times d^{-1}$
Re	Reynolds number, –
Re_{imp}	Reynolds impeller number, –
S	Covariance matrix, –
T	X-scores, –
T	Bioreactor diameter, m
t_a	Score vector corresponding to the ath principal component, –
TP_k^{max}	Peak concentration of the transport protein k, $mol \times L^{-1}$
TP_k	Transport protein k, –
U	Y-scores, –
U	Matrix of the left singular vector, –
V	Matrix of the right singular vector, –
V_{Bleed}	Bleed volume, L
$V_{Exchange}$	Exchange volume, L
$V_{Harvest}$	Harvest volume, L
$v_{i,j}$	Stoichiometric coefficient of species i in reaction j, –

List of Symbols

v_i	Stoichiometric coefficent of species i, $-$
$v_{NS,k}$	Incorporation rate of a nucleotide sugar, $mol \times s^{-1}$
V_R	Reactor volume, L
V_{Sep}	Volume of separation device, L
V_S	Gas superficial velocity, $m \times s^{-1}$
V_{tot}	Total volume, L
W^*	Adjusted weights of matrix X, $-$
X	Data matrix, $-$
X_d	Dead cell density, $10^6 \ cells \times mL^{-1}$
X_l	Lysed cell density, $10^6 \ cells \times mL^{-1}$
x_m	Vector of x observation of mth variable, $-$
$X_{V,meas}$	Measured cell density, $10^6 \ cells \times mL^{-1}$
$X_{V,SP}$	Cell density set-point, $10^6 \ cells \times mL^{-1}$
$X_{V,target}$	Cell density target, $10^6 \ cells \times mL^{-1}$
X_V	Cell density, $10^6 \ cells \times mL^{-1}$
Y	Yield, $\%$
$Y_{\mu,AMM}$	Ammonia-growth-dependent yield coefficient, $1 \times mol^{-1}$
$Y_{NS,k}$	Nucleotide-sugar-growth-dependent yield coefficient, $1 \times mol^{-1}$
$y_{test,i}$	y-value of the ith observation in the external set, $-$
z_j^{max}	Localisation of the peak concentration of E_j, $-$
z_k^{max}	Localisation of the peak concentration of TP_k, $-$
CO_2	Carbon dioxyde
H_2O	Dihydrogen monoxyde
HCO^-_3	Bicarbonate ion
K	Potassium
Na	Sodium
$NaHCO_3$	Sodium bicarbonate
O_2	Dioxygen
OH^-	Hydroxyde ion

1

Perfusion Mammalian Cell Culture for Recombinant Protein Manufacturing

In this introductory chapter, we discuss mammalian cell culture for the production of therapeutic proteins in the broader context of biotechnology and in particular of the biopharmaceutical industry. We begin with a short retrospect on the history of cell cultures for bioproduct manufacturing and eventually introduce recombinant technology, in order to appreciate how the present standards were established. An overview of the current market on recombinant therapeutic proteins provides some important understanding of the industry challenges to come and the contribution that continuous manufacturing can provide. We then introduce the various bioreactor types that can be used in this area and indicate the challenges that are faced in their development, design and operation.

The objective here is to put all these aspects in the right perspective and direct the reader to the chapters where each of them is specifically treated this book.

1.1 BIOTECHNOLOGY: FROM EARLY APPLICATIONS TO BIOTHERAPEUTICS

In general, biotechnology refers to any activity that uses living systems to create value through services or goods. The immense diversity of applications increases continuously with new discoveries in fundamental biology and, more recently, also with advances in computational sciences that allow extracting new information from the incredible amount of data generated with increasingly sophisticated analytical techniques.

Therapeutic proteins constitute a particularly important and lively part of this discipline. The number of indications treated with therapeutic proteins is increasing at a fast pace, which is very encouraging for patients suffering from various diseases – in particular, of autoimmune and oncologic nature. On the other hand, this represents an enormous market with sales in the order of billions of dollars each year, where manufacturing technologies play a fundamental role (Ecker et al. 2015, Walsh 2014, Zhang et al. 2016). Fermentation is probably one of the most common of such technologies and appears in a variety of complex industrial applications. It is probably more convenient to approach its description from an historical prospective, which is rooted in food processing.

1.1.1 First Applications of Fermentation

Humans began exploring biotechnology thousands of years ago, long before the existence of microorganism, genes or deoxyribonucleic acid (DNA) could even be imagined. However, progress in the area, as it is often the case, did not proceed continuously but through distinct discontinuities. Sometimes knowledge was stagnant for decades or even centuries until some technological breakthrough moved it to the next phase, through a sequence of events eventually leading to the modern context of biotechnology and biomanufacturing. The first and by far the longest phase started with food and beverage fermentations using mixed cultures of organisms whose identity remained unknown at least until the nineteenth century. Their primary function was to improve the preservation of aliments but eventually turned into the production of more sophisticated products. Cheese-making began about 8,000 years ago using microorganisms such as bacteria, yeasts and fungi to transform milk into a wonderfully flavoured but, most importantly, conservable product (Beresford et al. 2001). Fermentations of meat and vegetables also started early on and are still in use today not only for conservation purposes but also to create flavours that are part of the culinary legacy of our ancestors.

Fermented beverages could be traced back even earlier to the seventh millennia B.C. (McGovern et al. 2004). Wine and beer are part of the most ancient fermentation processes and are still today produced in large amounts, with all the positive and negative implications this has had on societies ever since. The oldest evidence of advanced beer brewing was traced back to around 5,000 years ago in China (Wang et al. 2016), while evidence for wine fermentation is traced to about 6,000 years ago, in today's Iran (McGovern et al. 1996, Soleas et al. 1997). These processes are currently well characterised, and methods for continuous beer production are being developed (Brányik et al. 2005, Pilkington et al. 1998).

It was only during the nineteenth century that fermentation was correlated to the activity of living organisms (Manchester 2007, Pasteur 1885). During the industrial revolution and with the advances in microbiology, these processes started to be understood and controlled (Caplice & Fitzgerald 1999).

1.1.2 Extraction of Valuable Metabolites and Proteins

The next technological milestone was related to purified monocultures of selected microorganisms used for the production of primary metabolites. The pioneer of these processes was probably Chaim Weizmann with his studies on the fermentation of acetone, butanol and ethanol in the 1910s. World War I motivated the development of fermentation technologies because of the large use of acetone and glycerol in munition production (Demain 2007, Wang et al. 2001). Many products were then synthesised using microorganisms including amino acids, vitamins, nucleotides, solvents and organic acids (Demain 2007, D'Este et al. 2017, Sano 2009, Stahmann et al. 2000, Zhang et al. 2016). The application of fermentation for the production of secondary metabolites can be regarded as a new technological breakthrough – which allowed the production of a new series of products such as toxins, pesticides and animal or plant growth factors and, of course, antibiotics, which alone had an unmeasurable impact on mankind since their discovery in 1929 by Alexander Fleming (Fleming 1929).

1.1.3 Recombinant DNA and the Manufacturing of Recombinant Proteins

Until this point, biotechnology was using existing organisms as such, in a controlled environment, with the purpose of collecting some selected secreted components.

Cultivation methods for industrialisation were developed mainly with microbial systems. The next and probably most significant breakthrough came through the so-called recombinant DNA technology developed after some pioneering work of three scientists – Paul Berg, Herbert Boyer and Stanley Cohen – in the 1970s. Actually, it was not so much the technique itself that was revolutionary but, rather, its application, which changed biology from a largely analytical science into a synthetic one (Berg et al. 2007). At that time, the access to therapeutic proteins was extremely limited and of animal origin. For example, insulin was of bovine or porcine origin and had significant adverse effects on the patients. With recombinant technologies, the human form of insulin could be produced at much higher safety and efficiency levels. The possibility to synthesise, multiply and integrate the DNA in a different expression system was revolutionary and had a terrific impact on medicine and many fields in biology, to the extent that it even changed the fundamental concepts of intellectual property for biological inventions (Berg & Mertz 2010, Cohen et al. 1973, Hughes 2001). Figure 1.1 shows Herbert Boyer on the *Time* magazine cover (next to one of the first appearances of the future Princess Diana). He co-founded Genentech (standing for **GEN**etic **EN**gineering **TECH**nologies); which, even today is one of the major companies in the biopharma world. The development of the technology for synthesising recombinant DNA is summarised through the major corresponding milestones in Table 1.1.

Recombinant technology is used for the production of highly specific therapeutic proteins. The DNA sequence coding for the protein of interest is inserted into a living host that is able to translate the DNA sequence into an amino acid sequence. When dividing, the host propagates the DNA information, or sequence, and synthesises the protein using its innate machinery (Colosimo et al. 2000). As mentioned earlier, microorganisms such as bacteria, yeasts, fungi, insects, plants and mammalian cells were used for a long time in this industry, but the development of recombinant DNA technologies opened the doors to entirely new production horizons. Indeed, even today, only the complex machinery of a living organism is able to synthesise large proteins to achieve desired biological activities. However, we now have the choice of which host to use, and this is often a key and decisive question. Evolution offered us a diversity of organisms that all have the same DNA coding system for the assembly of amino acids but differ in terms of growth behaviour, post-translational modification (PTM) capabilities, secretion efficiency, etc. Indeed, after transcription (DNA into RNA) and translation (RNA into an amino acid sequence), the cell machinery takes care of the protein folding, the addition of different sugar molecules (glycosylation) and several other reactions known as PTMs (Beyer et al. 2018, Higel et al. 2016, Walsh 2010). These changes are regulated by a complex network of enzymes and take place in the endoplasmic reticulum (ER) and in the golgi apparatus, as schematically illustrated in Figure 1.2. As an example, glycosylation can be described as a sequential processing of an oligosaccharide that is added to the protein structure in the ER. These added glycans are known to impact not only the stability but also the biological activity of the protein in the human body (Frenzel et al. 2013; Hossler et al. 2009; Jefferis 2005, 2009). A schematic view on how these transformations develop is shown in Figure 1.3. Starting from the upper left part, it is seen that the oligosaccharide chain is attached to the protein backbone in the ER. Then, some enzymatic reactions transform this glycan into oligomannose form, which is transferred to the Golgi apparatus – where, through other enzymatic reactions, different saccharides are removed or added. This complex

Table 1.1 Major events in the 1970s that are related to the development of recombinant DNA production and cloning methods © Republished with permission of Genetics Society of America, from personal reflections on the origins and emergence of recombinant DNA technology', P. Berg and J. Mertz, 184, 2010; permission conveyed through Copyright Clearance Center, Inc.

Year	Event
1969–1970	P. Berg and Lobban independently conceive ideas for generating recombinant DNAs *in vitro* and using them for cloning, propagating and expressing genes across species.
1971	Berg (1974) isolate the first plasmid bacterial cloning vector, *dvgal* 120.
1971	Robert Pollack first raise the concern regarding potential biohazards of cloning.
1971–1972	Jackson et al. (1972) and Lobban and Kaiser (Lobban 1972, Lobban & Kaiser 1973) concurrently and collaboratively develop the terminal transferase tailing method for joining together DNAs *in vitro*.
1972	Jackson et al. (1972) create first chimeric DNA *in vitro*.
1972	Mertz and Davis (1972) discover that cleavage with EcoRI generates cohesive ends. They use EcoRI plus DNA ligase to generate SV40 – *dvgal* 120 chimeric DNAs *in vitro*.
1972–1973	Cohen et al. (1972) isolate the drug-scalable bacterial cloning vector, psC101. They use it to construct, clone and express bacterial intra- (1973) and inter-species recombinant DNAs.
1973	Morrow et al. (1974) clone and propagate ribosomal DNA genes from a eukaryote in *E. coli*.
1973–1976	Renewed concerns regarding potential biohazards of cloning recombinant DNAs (Berg 1974, Berg et al. 1975, Singer & Soll 1973) lead to NIH Guidelines.
1974–1975	The initial Standford University/University of California, San Francisco (UCSF) (Cohen/Boyer) patent application relating to recombinant DNA is filed.
1976	Boyer and Swanson co-found Genentech, the first biotechnology company.
1980	Stanford/UCSF (Cohen/Boyer) patent issued by U.S. Patent Office.

reaction network can lead to a wide variety of different glycan species, which can be identified as more or less specific structures. It is very common to distinguish among different galactosylated forms, which include glycans that have one or more galactose attached. In this book, we distinguish among groups that are not galactosylated (G0) or have one (G1) or two (G2) molecules of galactose attached, as well as high mannose (HM) forms, referring to molecules exhibiting five or more mannose.

All these modifications play a fundamental role in the protein functionality and stability and are therefore of the greatest importance. This is the reason why mammalian cells are widely used in the production of therapeutic proteins, and this is the host we will focus on in the following. Although their cultivation is more complex and less effective in terms of productivity compared to bacterial or even yeast systems, their capacity to modulate certain post-translational modifications – like glycosylation, for example – makes them a powerful host for the production of recombinant proteins that are biologically active and safe for humans. In the next chapter, we discuss existing cultivation methods for mammalian cells.

Figure 1.1 *Time* magazine cover (1981) showing Herbert Boyer, Professor of Microbiology and Biochemistry at the University of California in San Fransisco and co-founder of Genentech (Hughes 2001). From TIME. © 1981, TIME USA LLC. All rights reserved. Used under licence

1.1.4 Product Quality and Quality by Design

Similar to any other industrial product, the quality of therapeutic proteins can be assesed through a list of physico-chemical characteristics, referred to as critical quality attributes, that have to fall within strict specifications. These are defined and

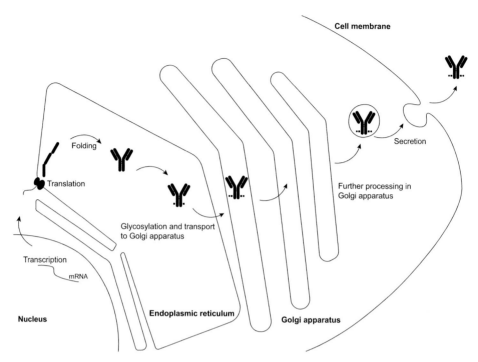

Figure 1.2 Protein pathway through the cell before secretion. After translation, the protein is folded in the ER, and some post-translational changes start to happen. The protein then travels through the Golgi apparatus, where further glycosylation transformations are regulated. The protein is then secreted to its functional location (Hu 2012).

controlled by the regulatory authorities in order to guarantee safety and efficacy of the drug. Opposite to chemically synthesised small-molecule drugs, which can be univocally defined by a single molecular structure, biological products, because of their size and complexity, are actually constituted by a heterogeneous population of very similar variants or isoforms. The sequence of assembled amino acids that represents the protein primary structure is controlled, but the host cell line and the process to express the protein have a direct impact on the PTMs – such as glycosylation or undesired chemical transformations like deamidations, oxidations or others which lead to undesired species or impurities. A biological product is therefore defined also by the corresponding production process, and its structure is defined based on the distribution of the different quality attributes. This is done through acceptable ranges that are more or less narrow depending on each attribute. One of the consequences is the difficulty of stating that two bioproducts are identical, which makes it intrinsically impossible to copy an existing biological drug. For this, the notion of generic that is used in the small-molecule pharma industry does not apply to the biopharma world. The concept of biosimilar was therefore developed to cope with the heterogeneity of the products in this industry. Guidelines were given by the authorities to define biosimilars and help manufacturers developing strategies for their production (Brühlmann et al. 2015).

In addition to the heterogeneity mentioned earlier, critical quality attributes (CQAs) include impurities which derive from the protein production process. Examples include DNA and host cell proteins (HCP) coming from the metabolism

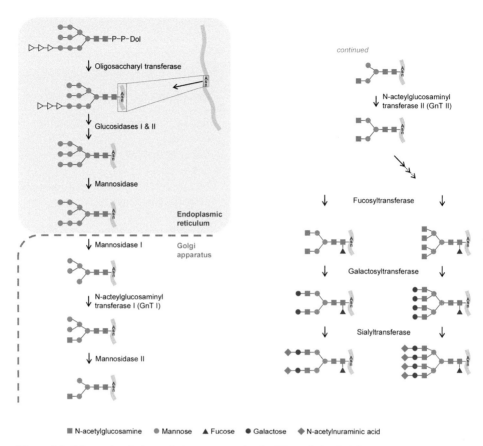

Figure 1.3 Schematic of glycosylation patterns in the ER and the Golgi apparatus. © Reprinted from *Biotechnology Progress*, David Brühlmann, Martin Jordan, Jürgen Hemberger, Markus Sauer, Matthieu Stettler and Hervé Broly, 31/3, 'Tailoring recombinant protein quality by rational media design', 2015, with permission from John Wiley and Sons.

of the host cells or low molecular weight (LMW) species and high molecular weight (HMW) species forms arising from protein fragmentation and aggregation processes, respectively. In general, all impurities are classified as product or process related, as shown in Table 1.2, and need to be properly taken care of in either the upstream or the downstream part of the production process. This requires a delicate coordination between the two because, as we will discuss later, it may be easier, if at all possible, to satisfy specifications for one specific critical quality attribute in one or the other of the two parts of the process.

The heterogeneity and variability of these attributes depends, on the one hand, on the given expression system and protein of interest, but also on the process used for the production of the molecule. Within product quality considerations, the concept of quality by design (QbD) has been strongly promoted in recent years (Woodcock 2014). QbD proposes a shift towards a better understanding of the product, its manufacturing process and the identification of potential critical product quality attributes (Rathore & Winkle 2009, Yu et al. 2015). The goal of QbD is to understand how these parameters are related to the process parameters and to develop quantitative relations among the two so as to design product quality based on process operating

Table 1.2 Non-exhaustive list and description of some typical product CQAs and some examples of CPPs that could affect them.

Critical quality attribute	Critical process parameter	Type	Description
Host cell proteins	Viability and cell density	Process	Impurity that may trigger immunogenic responses.
DNA	Viability and cell density	Process	Impurity that may trigger immunogenic responses.
Aggregates (HMW)	Titre and process duration	Product	Impurity that leads to loss of function and potentially in immunogenic responses.
Fragments (LMW)	Titre and viability	Product	Impurity resulting from chemical or proteolitic degradation; it can lead to loss of function and potentially immunogenic responses.
Charge variant	pH and temperature	Product	The charge distribution on a protein is often heterogeneous and impacts structure, stability, binding affinity, etc. Chemical degradations can modify the charge distribution and generate acidic or basic variants.
Glycosylation	pH and metabolite concentrations	Product	The glycoform distribution is characterised by various lumped forms: G0, G1, G2, sialylation, fucosylation, high mannose forms etc. Their proportions impact the biologic activity of the molecule.

conditions. The objective is to control, through the product quality attributes, biological activity, pharmaco-kinetics and pharmacodynamics, as well as safety and toxicity, in order to produce consistently safe and therapeutically efficient drugs (Rathore 2009, Rathore & Winkle 2009, Rouiller et al. 2012).

The quality of the product is, of course, strongly affected by the manufacturing process, both upstream and downstream. Critical process parameters (CPPs) are identified as acting on CQAs and must therefore be closely monitored. Downstream processing is typically regarded as the step where some undesired features of the product leaving the bioreactor can be modified to satisfy specifications, but this is not always possible. Indeed, downstream processes, whether in batch or continuous mode, are capable to remove or reduce within specifications process-related impurities, such as DNA and host cell proteins, as well as product-related impurities, such as aggregates and fragments (Steinebach et al. 2016b, 2017). Although more difficult, downstream processes – in particular, the multicolumn countercurrent solvent gradient purification (MCSGP) process – can change the charge variant patterns or isolate the main isoform at high purity and yield (Krättli et al. 2013). N-linked glycosylation isoforms, however, cannot be separated within the current downstream processes. Therefore, if needed, these patterns have to be modulated in the upstream process.

The use of continuous manufacturing, such as perfusion cell culture, impact the quality of the product in comparison to batch-like processes. This has relevant implications also at the monitoring level, for which regulatory environments already exist (Allison et al. 2015).

1.2 MAMMALIAN CELL CULTURES FOR RECOMBINANT PROTEINS

Mammalian cells naturally grow in the adherent state, and therefore, their cultivation requires suitable carriers. Adherent cells are still used in some well-established manufacturing processes, but growing cells in suspension, obtained through proper adaptation procedures, is much more efficient and is currently the industrial standard. For a given bioreactor, the density reached with suspended cells is, in fact, much higher than with adherent cells simply because there are no surface limitations.

Interestingly though, in the 1990s, perfusion processes were developed using adherent cells. Examples of commercial products include the recombinant follicle stimulating hormone (Gonal-f ©), interferon beta-1a (Rebif ©) and factor-VIII (Kogenate-FS ©), which are all large and labile molecules (Pollock et al. 2013). For this reason, they were cultivated in perfusion, and not batch or fed-batch mode, since labile molecules require low bioreactor residence times. In addition, adherent cells are intrinsically compatible with simple and efficient cell retention. In particular, with adherent systems, either fixed carriers, which prevented the cells from being harvested, or microcarriers, constituted by small floating beads of polymer whose density could be exploited in sedimentation devices to retain them inside the vessel, were used. The rare cells or cell debris present in the harvest stream were simply removed using depth filtration. In the case of cell suspension systems, only very low cell densities could be achieved, thus making cell settlers followed by depth filtration sufficient for reaching satisfactory cell retention inside the perfusion bioreactor.

However, at that time, industry largely favoured batch-like operations. From the 1990s to today, in fact, technologies around expression systems, media and process control went through enormous improvements (Chu & Robinson 2001, Lim et al. 2010). The adaptation of cell lines to suspension cultures offered new possibilities for manufacturing sufficiently stable proteins. Batch operations with suspension cultures were favoured in industry because of the increased productivity potential related to the higher cell concentration that could be reached compared to adherent systems. Reliable retention devices for even modestly concentrated suspension cultures were not available at the time. From the 1990s to today, the productivity or yield of this type of processes doubled every five years (Langer 2014, Meuwly et al. 2006). On the other hand, continuous cell culture was also used, but with some limitations. Low cell densities were achieved in perfusion due to the surface limitation for adherent cells, or to the retention capacity for suspended cells. Also, the number of stable molecules such as monoclonal antibodies (mAbs) increased in the company pipelines. This also motivated the development of batch systems, since low residence time was not a priority anymore for these molecules. The consequence is that today the industry standard for commercial manufacturing is fed-batch in stirred tank reactors of up to 25 kL (Moyle 2017). These reactors offer large mixing and mass transfer rates and can be used to cultivate many different cell types using different operation modes (Rodrigues et al. 2010). The downside is the large capital investment they require – which, according to the Biophorum Operations Group (BPOG), amounts to up to hundreds of millions dollars at commercial scale for a line of four bioreactors (Sawyer et al. 2017a).

Current market trends have also been well summarised by the BPOG, and four major drivers have been identified the continuous growth of the biopharmaceutical market, the introduction of new product classes, the cost pressure coming from the payers and, finally, the uncertainty of approvals for the new products. These factors put leaders in a delicate situation in which not only uncertainties about sales but also manufacturing capacity needs are growing. The past decades were mainly dominated by some blockbusters for each company that were driving the capacity needs. With the coming changes, new strategies that require more flexibility are needed. Large stainless steel plants might not be the desired standard anymore, mainly due to the large capital investments and the long time (up to five years) required to be operational.

Continuous manufacturing is envisioned as one way to respond to the needs of these new market trends. If the production can be intensified, the bioreactor size, and its corresponding footprint, could be significantly reduced. Enabling technologies exist today for high cell density (up to 100×10^6 cells/mL and more) culture filtration with efficient cell retention (Barrett et al. 2018). This, together with the reduced lead time needed between the decision to build a commercial capacity and its first operation, promotes the introduction of continuous technologies. Single-use technologies are often mentioned in this context because they also offer an increased flexibility and a decreased capital investment.

In the following section, we briefly summarise the key aspects of the various bioreactor operating modes in order to put in a better prospective their capabilities and limitations.

1.2.1 Batch Operation

Batch operation is the simplest operation mode since it does not require any addition or removal of media during operation (Al-Rubeai 2015, Hu 2012, Zhou & Kantard-jieff 2014). The bioreactor is inoculated at a defined cell density, and then the cells grow, while the product is being secreted, up to some maximum level. Parameters like temperature, pH, mixing and dissolved oxygen are controlled in the culture vessel. The media used to start the batch bioreactor contains enough nutrients to sustain the culture for the entire batch time. Components typically include amino acids, vitamins, trace elements, glucose, growth factors and salts. In the earlier times, these culture media also contained some supplements extracted from living beings such as serum of animal origin or peptones issued from yeasts or plants. These supplements were critical for safety and consistency reasons (lot-to-lot variations); therefore, media containing only known concentrations of defined components – usually referred to as chemically defined (CD) media – were developed. In all these cases, the media are constituted by a rather complex mixture of many components. In batch mode, the nutrient content is limiting and limited – limiting because once the nutrients are consumed, the cell culture is not sustained anymore, and limited not only because of solubility or stability constraints of each component but also because the starting condition must be in an acceptable pH and osmolality range in which the cells can grow. This implies that the concentrations cannot simply be increased indefinitely to increase the media capacity to grow cells.

Another limiting factor for the cell culture is the accumulation of toxic metabolites. During the culture, the cells secrete a diversity of metabolites that are known to be detrimental to cell health if too concentrated. Some of these metabolites like

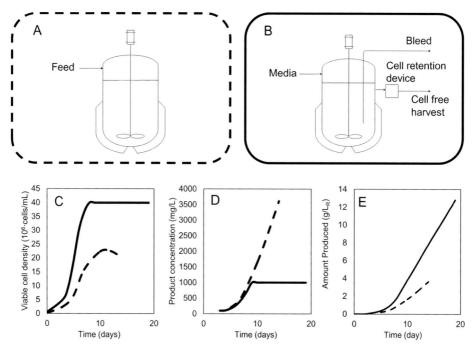

Figure 1.4 Schematic of fed-batch (A) and perfusion (B) culture of a recombinant cell line. In a fed-batch bioreactor, feeds are added during the run, and the product accumulates in the bioreactor. In perfusion, media is continuously fed, and harvest is continuously removed. The cell retention device clarifies the harvest stream, and the product is (in this example) not retained and, therefore, not accumulating. The bleed stream is used to remove excess cells and keep a steady concentration, in this example, at 40×10^6 cells/mL. The schematic depicts cell density (C), protein concentration inside the bioreactor (D) and cumulative specific production (E) as a function of time; broken lines represent fed-batch and solid lines perfusion. © Reprinted from *Biotechnology Advances*, 36/4, Jean-Marc Bielser, Moritz Wolf, Jonathan Souquet, Hervé Broly and Massimo Morbidelli, 'Perfusion mammalian cell culture for recombinant protein manufacturing: A critical review', 2018, with permission from Elsevier.

ammonium or lactate are well known and can be monitored, but others, not well characterised, may also interfere with the culture.

1.2.2 Fed-Batch Operation

The operation of a fed-batch process is similar to a batch operation, but concentrated nutrient solutions (feeds) are added to the culture during operation to avoid the depletion of essential nutrients, as shown in Figure 1.4. This strategy enables the increase of the maximum cell density and also the culture duration with respect to batch operations. Various feed addition strategies are possible: it can be bolus or continuous addition, and the added volumes can be based either on an established platform or on some reference nutrient concentration. Several different feeds are often used, for example, to control the glucose concentration or avoid the depletion of specific nutrients. Some components that are unstable or difficult to solubilise might also require an additional feed stream. However, one issue is not resolved by this strategy: the accumulation of waste and possibly toxic metabolites in the bioreactor, which imposes in all cases an upper bound to the maximum achievable viable cell density.

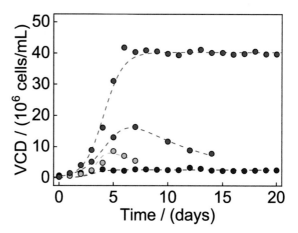

Figure 1.5 Viable cell density as a function of time for batch (green), fed-batch (blue), chemostat (purple) and perfusion (red) bioreactors.

1.2.3 Perfusion Operation

Continuous cell culture operation is possible with a constant media addition – which, in order to keep constant the bioreactor volume, has to be coupled with a constant harvest flow. If no device is used to retain the cells in the bioreactor, this operation is referred to as chemostat. However, since the growth kinetics of the cell culture – especially with mammalian cells – is rather slow, it is preferred to retain the cells inside the bioreactor, which corresponds to the perfusion operation mode. One exception that needs to be mentioned here is the case where the separation device also retains the product of interest inside the bioreactor. This is a relatively rare situation which is not further considered in the following paragraphs (Yang et al. 2016).

A perfusion bioreactor is characterised by two outlet streams as shown in Figure 1.4: the harvest, which contains no cell and controls the concentration and residence time of the secreted product, and the bleed (containing also the cells), which is used to control the cell density in the bioreactor and is obviously significantly smaller. In this operation mode, the toxic metabolites are removed from the bioreactor while cells are retained, thus enabling maximum viable cell density values larger than in all other operations to be reached. This is illustrated in Figure 1.5, where the four operating modes are compared for a representative case. On the other hand, the nutrients and secreted product are also removed from the reactor, thus leading to lower titres and higher media-specific consumption. The various aspects relative to the operation, monitoring and control of perfusion bioreactors are discussed in Chapter 2.

Perfusion operation, as well as the chemostat, differs significantly from batch and fed-batch because they are open or continuous systems, with constant inlet and outlet flows, which after a sufficient transient time tend to balance each other and reach a time-independent state usually referred to as steady state. This is well described by the classical ideal model usually referred to as CSTR (Fogler 2008). A qualitative comparison of fed-batch and perfusion operations is shown in Figure 1.4C, D and E with respect to viable cell density, titre and cumulative production.

Because the maximum viable cell density that can be reached using perfusion is significantly higher than in batch or fed-batch operations, the limitations in terms of cell density and culture duration shift from nutrient depletion or toxic compounds accumulation to cell retention, oxygen supply and cell stability. The challenges become to control the flow of the different streams, to guarantee the proper functionality of the retention device and to maximise the productivity with a minimal amount of media consumption. To achieve these goals, solid process development and operation strategies are required.

1.3 DEVELOPMENT AND OPERATION OF PERFUSION CULTIVATIONS

The development of a perfusion process consists in the design of all the operating conditions at which the process has to be run. Some of these are not significantly impacted by the continuous operation mode. This is the case, for example, of the bioreactor environment, which includes temperature, pH, mixing, oxygen concentration, etc., where the same concepts as for batch operation can be applied. However, there are other variables, such as the various flow streams (media feed, harvest, bleed), that need to be properly defined and controlled to achieve and maintain a steady-state operation. The cell retention device needs to be carefully designed to assure stable operation of the bioreactor for the entire duration of the process. Also, media design and cell line selection need to be adapted with respect to batch operation in order to better fulfil the requirements of the continuous operation. These aspects are briefly discussed in the following section in order to draw a global picture of the complex task of designing perfusion bioreactors across the various relevant scales, while a detailed and quantitative analysis is presented in the rest of the book, and particularly in Chapter 4. On the other hand, the different cultivation scales, including scale-down models, that are used to define these operating conditions are described in Chapter 3.

1.3.1 Cell Retention Device

A variety of different cell retention systems has been described in the literature, which are based on differences in either size or density to separate cells or microcarriers from the liquid harvest flow (Bielser et al. 2018, Pollock et al. 2013, Voisard et al. 2003). In fixed-bed bioreactors, the cells are attached to some packed carriers which are kept inside the reactor. Microcarriers are instead constituted by small beads dispersed in the suspension and kept inside the bioreactor by appropriate spin filters or settlers.

For cells in suspension, the separation is more difficult because of their reduced size and density differences compared to the microcarrier beads. Partial retention can be achieved with well-designed gravimetric settlers, but the separation yield decreases as the cell density increases and more performant systems are typically required. Tangential filtration is based on small pore sizes that retain mammalian cells and a tangential flow to alleviate membrane fouling. Various commercial solutions based on this principle exist today and are discussed in Chapter 2.

1.3.2 Process Monitoring and Control

In order to properly operate a perfusion bioreactor, we need to control three liquid flowrates: the media inlet flow (Q_{feed}), also referred to as the perfusion rate; the

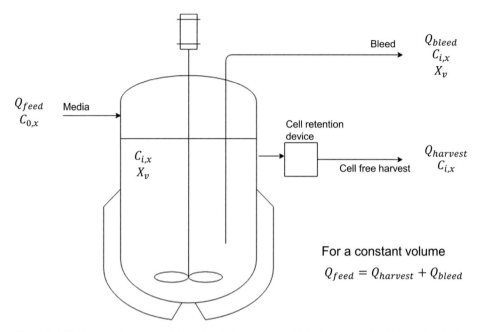

Figure 1.6 Perfusion bioreactor with media feed, harvest and bleed streams. A cell retention device is present on the harvest line. © Reprinted from *Biotechnology Advances*, 36/4, Jean-Marc Bielser, Moritz Wolf, Jonathan Souquet, Hervé Broly and Massimo Morbidelli, 'Perfusion mammalian cell culture for recombinant protein manufacturing: A critical review', 2018, with permission from Elsevier.

harvest flow ($Q_{harvest}$) and the bleed flow (Q_{bleed}), as shown in Figure 1.6. In order to maintain a constant volume in the vessel, the inlet flowrate must obviously be equal to the sum of the two outlet ones. These flows can be controlled using different strategies; for example, one can fix either the feed or the harvest flow as being constant and adjust the other two accordingly. The bleed flowrate is typically manipulated in order to control the cell density, while the remaining flowrate – i.e., feed or harvest – is manipulated in order to keep constant the weight of the bioreactor and, consequently, the culture volume.

In order to keep constant or control the flowrate of a liquid stream, one can envision different approaches such as to use a calibrated pump, a non-invasive flow meter based on ultrasounds or a gravimetric flow controller. All options are viable and mostly depend on available equipment. The easiest to implement is, of course, a pump with a defined flowrate, such as a volumetric pump, but this might not be the most precise or reproducible. Therefore, flowrate controllers using a feedback loop based on a measurable physical parameter such as weight or liquid velocity are more recommendable.

The bleed stream is ideally continuous and can be controlled offline, using a pump with a daily flowrate adjustment based on an external cell count, or online, using a capacitance probe that is able to monitor the biomass. A close control of the cell density in the bioreactor is crucial to maintain a stable steady-state operation.

Thus summarising, three streams and one retention device are the extra requirements for perfusion cell cultures compared to fed-batch operations. Technologies are available for their realisation across various scales, from benchtop to manufacturing,

Table 1.3 Non-exhaustive list of small scale cell culture devices, with comparison of control strategies and operation modes. © Reprinted from *Biotechnology Advances*, 36/4, Jean-Marc Bielser, Moritz Wolf, Jonathan Souquet, Hervé Broly and Massimo Morbidelli, 'Perfusion mammalian cell culture for recombinant protein manufacturing: A critical review', 2018, with permission from Elsevier.

Parameters	Deepwell plate	Shake tubes	Microscale	Lab-scale
Mixing	Orbital shaking	Orbital shaking	Stirred	Stirred
pH control	No	No	Yes	Yes
DO control	No	No	Yes	Yes
pCO_2 control	No	No	Yes	Yes
Aeration	Passive	Passive	On demand	On demand
Cell retention	Centrifugation or Sedimentation	Centrifugation or Sedimentation	Sedimentation	ATF, TFF or other
Perfusion operation mode	Semi-continuous	Semi-continuous	Semi-continuous	Continuous

but the use of cell retention devices is not always straightforward. In particular, the smallest scale may be challenging and requires proper scale-down models.

1.3.3 Scale-Down Models

It is good practice in engineering to develop processes at a small scale and, once the preferred operating conditions are identified, to transfer them to the commercial scale. This is especially true for biopharma processes due to the very high cost and time requirement of any experimental activity. The problem is that continuous flows and cell retention, which are essential features of perfusion cell culture, are difficult to scale down for technical reasons. Batch-like technologies benefit from established scale-down models such as the deepwell plate (DWP), the shake tube (ST) or the micro-scale bioreactor (msBR). Actions like sampling for offline measurements, also referred to as in-process control (IPC), and feeding can, in fact, be easily mimicked in these systems. Strategies for pH and dissolved oxygen (DO) control have also been established using, for example, incubation in a controlled environment and agitation to create mixing and promote interphase oxygen and carbon dioxide transfer in the culture. The characteristics of the various devices available on the market are summarised in Table 1.3.

As mentioned earlier, the case of perfusion is more complex, and this can be mimicked using semi-continuous operations. For this, cells are periodically separated from the fermentation suspension either by centrifugation or sedimentation. The cell-free medium is then harvested and replaced by fresh medium, as shown in Figure 1.7.

Still in the area of scale-down models, but for a bit larger volumes, a device that enables to control up to 24 bioreactors with a culture volume of 250 mL each has been recently commerialised (Zoro & Tait 2017). This device allows the

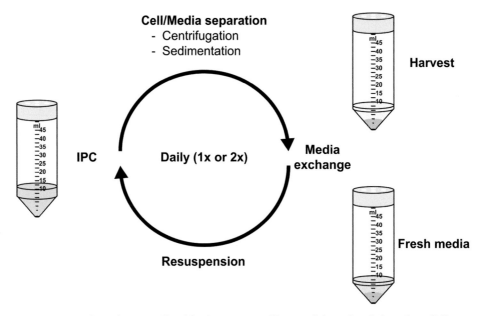

Figure 1.7 Semi-continuous cell cultivation strategy illustrated here for shake tubes. Cells are separated using either centrifugation or sedimentation. Then, the spent media is exchanged with a defined volume of fresh media, and the cells are resuspended. © Reprinted from *Biotechnology Progress*, Jean-Marc Bielser, Jakub Domaradzki, Jonathan Souquet, Hervé Broly and Massimo Morbidelli, 'Semi-continuous scale-down models for clone and operating parameter screening in perfusion bioreactors', 2019, with permission from John Wiley and Sons.

implementation of both continuous operation and cell retention. The system offers two filtration patterns: tangential flow filtration (TFF) and alternating tangential flow (ATF) filtration, in addition to online control of pH and DO, at-line control of cell density, and constant media and harvest flow control. There is also a single-use equipment provider (PerfuseCell ©) that has a large choice of different lab-scale bioreactor set-ups, including different types of cell retention devices. This allows for customer-designed bioreactors – including vessel size, number of lines, agitator shape, sparger and even different types of single-use and pre-sterilised retention devices. Another report mentions the development of perfusion capacities at 250 mL scale using modified bioreactor systems (Chotteau 2017). Although more similar equipment is expected to become available soon on the market, those based on semi-continuous operation will remain a valid option due to their low cost and ease of operation. The different characteristics of these approaches and the reliability of scale-down models in the sub-litre range are discussed in Chapters 3 and 4.

1.3.4 Media and Cell Line Development

Ad-hoc media development is a crucial aspect for perfusion processes since it represents a strong component of the production cost. Due to the outlet flow, significant amounts of nutrients, in fact, leave the bioreactor non-utilised. Consequently, the amount of media consumed per mass of product tends to be larger than in batch-type processes. This is a crucial point in the development of perfusion bioreactors that will be discussed in Chapters 2 and 4.

Another key element of recombinant technology is the development of appropriate cell lines. As mentioned earlier, this is based on the selection of a host cell line and a defined DNA sequence encoding for the protein of interest, which we assume given at this point. For mammalian cells, transient and stable transfection of the given DNA can be performed. Transient transfection is very fast, and the DNA is not integrated into the host genome and, therefore, is not preserved, or copied, over subsequent generations. Nevertheless, for a short period of time the strand will be expressed. This production method is typically used in drug discovery and in the very first development stages when low amounts of product are required and the quality constraints are rather loose (Pham et al. 2006).

However, in order to establish a manufacturing cell line, the transfection must be stable, and the clonality of the cell line must be proven. Different transfection methods exist with mostly random integration (location and number of copies are not controlled) of the vector into the genome. This results in a very large pool of different clones. For transient expression, this pool would be used as such, but for stable transfection, some selection pressure has to be applied in order to identify the rare clones that integrated the transfected DNA into their genome, and thus replicated it with each cell division. The next step is to split these cells individually. Different methods ranging from limiting dilutions to more complex technologies using flow cytometry are used for this separation. Well plates are commonly used to explore a large number of conditions. The screening procedure is based on the clone growth rate and productivity. These are used as indicators to select a number of them, scale them up and repeat the screening procedure with a narrower range of clones. This procedure is repeated until converging to a single clone which is then selected for manufacturing as schematically illustrated in Figure 1.8 (Wurm 2004).

It is to be noted that media design and clone screening occur somehow in parallel and, as the candidates become fewer, the experiments are conducted using larger volumes, thus moving from static plates to agitated deepwell plates, shake tubes, micro-scale bioreactors, lab-scale bioreactors and finally to the pilot and commercial scale (Böhm et al. 2004, Rouiller et al. 2016). Since it is expected that the best clones for operating in the batch mode are not necessarily the best ones in the perfusion mode, the scale-down models used in this procedure should be selected accordingly.

1.3.5 Scale-Up to Clinical and Commercial Production

The process development discussed in the previous sections ranges from the millilitre to the litre scale, which is the one typical of benchtop bioreactors. The next step is to scale the bioreactor up to the clinical and commercial scale.

When considering a traditional stainless steel three-phase bioreactor, this scale-up procedure involves the usual operating variables which are relevant in determining the performance of these reactors such as temperature, mixing, oxygen and carbon dioxide concentration and others, which are rather classical in the operation of three-phase reactors. A wide variety of methods, procedures and models are available for this in chemical reaction engineering and can be easily adapted to the case of bioreactors (Benz 2011, Shah 1979). This is discussed in Chapter 5, where we specifically focus on those aspects which are peculiar of continuous cell cultures. One of the major issues is related to the cell retention device. It is important to extract information from small-scale experiments, such as the maximum flowrate per surface area or the residence time of the cells inside the filtration device, for the appropriate

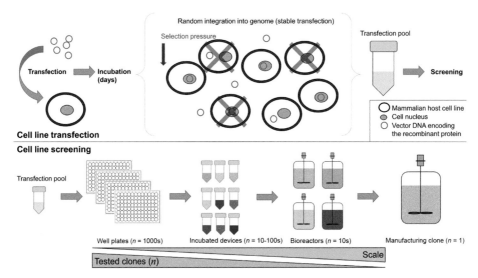

Figure 1.8 Manufacturing cell line development workflow. The cell line transfection step is a random insertion of vector DNA containing the encoding region for the target recombinant protein. Selection pressure is applied to the cell pool in order to favour the survival rate of the cells that integrated the vector. This heterogeneous pool of cells is cultivated for a short amount of time until viability recovers. The second step is the cell line screening. First, the cells from the transfection pool are split into well plates (target is one cell per well). This single cell by dividing gives rise to a clonal population. Each clone is tested in different scale-down platforms (see more details in Chapter 3). As the experimental scale increases, the number of clones decreases until the final choice of the manufacturing clone (Wurm 2004). Colours in the schematic shake tubes and bioreactors each represent a different clone.

design of large-scale units. The challenge is to ensure reliable and robust operation over the entire culture period, which implies sufficient cell retention with negligible product retention, low membrane fouling and appropriate average residence time of the cells in the external loop outside the bioreactor.

Another important trend which needs to be mentioned refers to single-use bioreactors. These technologies have a great potential with respect to decrease the investment costs and simplify unit operations, particularly with respect to cleaning and regeneration operations (Jacquemart et al. 2016). Indeed, several industrial operators are proceeding in this direction and more and more single-use units are becoming available on the market. Nevertheless, these systems are less well characterised than traditional stainless-steel vessels and questions regarding mixing and aeration are often difficult to answer properly. It is clear that we will see more development in this direction in the near future, although it is difficult at this point to predict to which extent they will eventually replace the traditional stainless-steel equipment (Sawyer et al. 2017a).

Finally, Chapter 6 discusses various mathematical models that can be used to describe different phenomenon. Classical models used in reaction engineering can be used to describe mass transport and chemical transformations occurring inside the bioreactor (Shah 1979). Computational fluid dynamics (CFD) is typically used to understand how the agitation impacts the homogeneity of the culture. This is an important feature that needs to be well described in lab-scale bioreactors and, more importantly, at the manufacturing scale. Other types of models can be used to describe the cell biology. For example, some metabolic pathways or some particular portion of the cell, such as the Golgi apparatus, can be modelled to understand the impact of changes in the reactor operating conditions. Each one of these – and, even more so, a proper combination of all of them specific for a given target – can provide an important contribution not only in the scale-up of bioreactors but also at the level of process design and development, as well as in understanding the impact of various operating conditions on specific product quality attributes.

1.4 CONCLUSION

Biopharmaceuticals – that is, biologically active proteins used to treat various types of diseases – represent today a well-established and relevant industrial sector. Manufacturing is a very important component of this industry, due to the relatively large production volumes, the high costs and the pressure coming from time and market needs. Regulatory constraints and the lack of understanding/control of the biological component are certainly major factors that refrain sometimes the implementation of the latest technologies. Nevertheless, other drivers like market changes and cost pressure are pushing industry to question its standards and will eventually push for change through innovation.

Perfusion is one of the technologies that could contribute to support such a change. This book is dedicated to perfusion bioreactors and is intended to provide the tools needed to understand their advantages and limitations so as to learn when to use them and how to design them to best exploit their potential across all scales from laboratory to commercial. Although already applied in the past, the current use of perfusion bioreactors in the context of process intensification and continuous manufacturing has a much larger scope which requires a better understanding. It is the scope of this book not only to underline the limitations of this technology but also to indicate the path for their removal or alleviation.

A large part of the book discusses the various aspects of process development of perfusion cell cultures. Some of these are common for all continuous industrial processes, like monitoring and control, while others, such as scale-down models, are particular to biopharmaceuticals. The steady-state concept, which is the basis for the operation of any continuous process, is discussed with some care, due to the presence of living entities which inevitably change their behaviour in time and consequently affect the bioreactor steady-state behaviour.

In the last part of this book, we address the role of large scale perfusion bioreactors in manufacturing biopharmaceuticals and their design and scale-up, also in combination with modelling techniques. This involves the important case of batch process intensification, where perfusion is used coupled to batch-type technologies in order to improve their productivity. On the other hand, perfusion bioreactors can be used alone in the upstream part of the production process, followed by the downstream section where the product is purified to specifications using batch- or continuous-type technologies (Pfister et al. 2018). In the second case, we can aim at an integrated continuous unit, where the up- and downstream are operated in a fully integrated manner, just like a single process. This represents probably the greatest innovation in this area. Unthinkable until a few years ago, it is becoming now reality and an attractive choice in a number of commercial scenarios.

2

Perfusion Bioreactors: The Set-Up and Process Characterisation

In this chapter, we give an overview of the challenges and objectives in operating mammalian cell perfusion cultures and provide guidelines for the design and set-up of lab-scale bioreactor systems. Next, we illustrate the control structure needed to maintain long-term stable and viable cultures and then provide a section on the media design. In the last part, we discuss steady-state operation, reactor and process dynamics, including product quality considerations. We also discuss comparisons with current technologies, particularly with respect to product quality.

2.1 CHALLENGES AND OBJECTIVES IN PERFUSION CULTURE OPERATION

Before entering the details of the realisation, control and operation of perfusion bioreactors at lab scale, it is convenient to put in perspective the challenges and the objectives connected to the application of these units.

2.1.1 Continuous Operation Mode

Depending on the specific application, continuous or perfusion bioreactors can enter in different ways in biomanufacturing. The combination of continuous nutrient supply and removal of the protein of interest and the toxic by-products coupled to cell retention inside the bioreactor allows high cell densities to be generated while maintaining the cells in a viable and growing state. In this context, we distinguish two different types of operation: on the one hand, continuous cultures without an additional cell-removing stream – the so-called bleed stream, typically applied for the generation of high cell density cultures – and on the other hand, continuous cultures with such an additional cell-removing stream, which enables long-term perfusion cultures and the achievement of steady-state conditions.

In the first case, continuous units are operated in seed trains, either for the generation of high density cell banks in perfused mode or for the operation of N-1 perfusion cultures, both targeting high cell density inoculum for either batch or continuous processes (Heidemann et al. 2010, Pohlscheidt et al. 2013, Yang et al. 2014). Another application is the intensification of batch bioreactors, the so-called concentrated fed-batch (CFB) operation. During a CFB, not only the cells are retained in the bioreactor but also the protein itself, while toxic by-products are removed in the harvest stream. This is implemented by using a filtration device with a smaller pore size – e.g., 50 kDa instead of 0.22 μm. In this case, the process is operated at a medium exchange rate enabling cell growth to much higher densities before reaching a medium limitation. The continuous operation intensifies the traditional batch-technology enabling up to a fivefold increase of cell density and harvest titre (Yang et al. 2016). Another application of continuous operation is the non-steady-state operation mode. The set-up is similar to a common perfusion process, but it is operated at a given medium exchange rate, continuously removing toxic by-products and the protein of interest. As cells are retained in the system, they grow until medium limitations are reached in terms of nutrient depletion at the given rate of medium exchange (Du et al. 2015, Gomez et al. 2017). All these systems are intended to increase the viable cell density and harvest yields compared to the traditional fed-batch processes.

The second type of operation of perfusion cultures targets long-term steady-state cultures, as in the context of the integrated continuous manufacturing of therapeutic proteins. In this case, it is not only necessary to guarantee accurate control of medium supply and perfusion rate and constant reactor volume but also to remove cells from the bioreactor in order to keep the viable cell density inside the bioreactor constant, as described in the previous chapter. To that end, cells are removed through the bleed stream, allowing the operation at constant cell density and, thus, at steady-state conditions.

It is worth pointing out that a continuous perfusion process typically consists of two different phases: an initial cell accumulation phase where sufficiently high cell numbers are generated for the next production phase, where the cell density is maintained stable at a constant level. This can be performed either in one reactor or split into an N-1 seed culture and an N-stage production reactor. During the first phase, high perfusion rates are chosen to favour cell growth, and during the production phase, perfusion is decreased and maintained at an optimal level for producing the target protein with minimum cell growth and medium consumption.

2.1.2 Challenges

In general, the operation of perfusion cultures requires solid design, robust units and a reliable control strategy of the bioreactor system. The addition of a cell retention device adds complexity and limitations to the whole system, which requires specific engineering development to deal with issues like aeration, carbon dioxide accumulation, mixing and homogenisation, fouling and failure of the cell retention device, process monitoring and control and so on. Operation at high cell densities up to 10^8 cells/mL leads, in fact, to higher overall supply and consumption rates of metabolites and gases, as well as higher viscosity values which – in combination with the shear sensitivity of mammalian cells – require specialised technological solutions.

In terms of gas supply and removal, this implies that the set-up has to support high volumetric gas–liquid mass transfer rates ($k_L a$) and volumetric gas flowrates for sufficient supply of oxygen and removal of carbon dioxide (Gray et al. 1996, Sieblist et al. 2011). For example, in order to achieve and maintain viable cell densities in the order of 10^8 cells/mL, large oxygen mass transfer coefficients in the order of 5–55 h^{-1} are necessary (Ozturk 1996), while carbon dioxide partial pressures should not exceed 100–150 mmHg (Kimura & Miller 1996, Pattison et al. 2000, Zhu et al. 2005).

A typical problem at high cell densities, particularly at larger scales, is to enable sufficient homogenisation of the cell culture broth. Insufficient homogenisation may lead to spatial gradients in both nutrient and gaseous compositions, which may affect the cellular metabolism and, thus, the achievable viable cell density. This requires a careful design of the mixing conditions – including the selection of the impeller, the baffle configuration and the stirring speed – to guarantee sufficient mixing without providing excessive stress that would harm the cell viability (Ozturk & Hu 2006).

A robust control strategy of the bioreactor conditions and culture-specific variables is important also to obtain stable perfusion cultures. This includes the tight, short-term control of the different inlet and outlet mass flowrates, viable cell density and reactor environment (pH, temperature, oxygen and carbon dioxide levels) while maintaining a constant reactor volume. Particularly relevant in these devices are long-term stability issues which relate both to mechanical devices, like preventing plugging due to fouling and protein retention in the cell retention device, as well as to the cell metabolism and the possible occurrence of ageing or stability issues (Ozturk 2014).

2.1.3 Objectives

Depending on the type of application, different requirements with respect to the bioreactor set-up, operation and control have to be considered. However, in general, the following three objectives need to be realised in order to fully exploit the potential of perfusion cultures.

The first and probably most important one is to run a stable process. This includes the mechanical and biological aspects described earlier. In the particular context of an integrated continuous process where the perfusion bioreactor is directly connected to the subsequent capture step, this becomes a prerequisite to allow stable operation of the entire downstream portion of the plant.

A second point is to achieve high productivity and medium utilisation during protein production, whether this refers to concentrated fed-batch, perfused non-steady-state culture or long-term steady-state operation. For this, the bioreactor operation has to be optimised in terms of high viable cell density, high titre and medium utilisation. In all applications, as well when generating high cell density banks and inoculum, it is important to achieve high cell-specific productivity values.

The third objective refers to the control and uniformity of product quality attributes. During steady-state perfusion cultures, the cells operate in an environment which is constant in time and, therefore, tend to maintain their metabolism unchanged and then always produce the same protein. Coupled to the constant and short product residence time within the reactor, final product quality patterns, including charge variant and glycosylation, are expected to be more uniform over time. Less exposure to protein-degrading enzymes due to shorter residence time is, in fact, expected to significantly reduce the probability of post-translational transformations such as protein fragmentation and aggregation, protein misfolding and as various types of degradation reactions – like deamidation, oxidation, sialylation, C-terminal lysine cleavage and isomerisation (Jordan & Jenkins 2007, Khawli et al. 2010). On the other hand, the presence of a continuous feed facilitates the introduction of suitable supplements in the bioreactor so as to suitably modify the product quality attributes.

The ability to properly design and develop perfusion bioreactors so as to deal with these objectives is the key to exploit the full potential of these units.

2.2   EQUIPMENT

Let us now consider the equipment necessary to develop and operate a lab-scale bioreactor system that meets the requirements for a stable perfusion culture. Besides the bioreactor system itself, equipped with proper control devices, this includes sensors for monitoring culture-specific parameters and suitable cell retention devices.

2.2.1 Bioreactor

Perfusion cultures can be operated in various different devices ranging from single-use units, like wavebag bioreactors (Clincke et al. 2013a, 2013b), to the classical stirred tank vessels (Karst et al. 2016, Xu & Chen 2016). For the purpose of running a continuous culture, the reactor system should be equipped with a number of connecting ports so as to accommodate all necessary tubes and fittings.

Whether operated in batch or continuous mode, bioreactors require reliable monitoring and control of various operating variables such as temperature, pH and dissolved oxygen (DO), through the proper operation of various pumps and mixing devices. In addition, the perfusion reactor system should enable the connection to medium, harvest and bleed bags or holding vessels, as well as the connection to a cell retention device that typically represents an additional part of the set-up. This can either be an external device or implemented inside the bioreactor itself (Zhang et al. 2015b). Besides the control units, the reactor has to be equipped with a stirring/mixing device, an aeration/sparger module, a condenser, sample and inoculation ports, connections for the addition of antifoam and base for pH control. Devices for aseptic connection are also necessary – e.g., a sterile welder and sealers.

For the configuration of the bioreactor, suitable geometrical parameters have to be chosen. In the case of a stirred vessel, these are particularly relevant at the large scale. Such parameters include the type of stirrers and baffles and all the relevant distances, such as the ratio of impeller and tank diameter or the impeller distance from the reactor bottom. A suitable sparger and stirrer that allow breakage and complete dispersion of gas bubbles at reasonable impeller speeds and elevated gas flowrates must also be chosen, without harming the cell viability. These aspects are discussed in Chapter 5, where suitable design criteria are also provided.

2.2.2 Sensors

The proper monitoring and control of the bioreactor operation necessitates several sensors. These include a pH probe, DO probe, temperature sensor, CO_2 sensor and biomass probe. The pH, DO and temperature probe represent industrial standards, often part of the bioreactor system modules, which are used to monitor and control the operation at predefined operating set points. Monitoring and control of CO_2 are important because high levels of dissolved CO_2 result in reduced cellular growth and productivity (Garnier et al. 1996, Kimura & Miller 1996). Biomass probes, usually based on capacitance measurements, are used for the online monitoring and control of the biomass inside the bioreactor that is often correlated to the viable cell density. In particular, biocapacitance has been shown to be linearly correlated to the viable cell density and, therefore, provides the ideal tool for monitoring and controlling this important process parameter (Carvell & Dowd 2006).

In addition, more complex sensors such as Raman, near infrared (NIR) or fluorescence sensors are and will be increasingly considered. These typically need to be coupled with suitable statistical multivariable data analysis techniques in order to estimate, from the measured complex spectra, process variables such as metabolite and protein concentrations or critical quality attributes, like N-linked glycosylation patterns or aggregates (Luttmann et al. 2012, Mercier et al. 2016, Rathore 2014, Rathore et al. 2018). However, applications are currently still limited to batch operation modes. The extension to longer operation times requires robust sensors with long lifetime, resistant to fouling and sterilisation procedures (e.g., autoclaving, gamma sterilisation). A more complete discussion of these and other monitoring devices is provided in Chapter 3.

2.2.3 Cell Retention Device

The cell retention device is a key element of perfusion cultures, and its longevity and scalability are fundamental for the long-term stability and reliability of the process. Cell retention systems are based on different physical methods for cell separation – in most cases, either size or density. The immobilisation and entrapment of cells in larger beads or fibres could, in principle, provide an effective method for retaining cells inside a bioreactor, but their operation and handling are often not convenient (Ozturk & Kompala 2006). In addition, there are also strong limitations in terms of maximum viable cell density, dictated by the limited availability of surface area compared to volume, like in suspension cultures. More suitable cell retention devices include cross-flow filtration systems, hollow-fibre modules, spin filters, gravity settlers, centrifuges and acoustic wave separators (Bielser et al. 2018, Bonham-Carter & Shevitz 2011, Voisard et al. 2003). Each of these shows advantages and drawbacks. Retention by size is typically used in filtration devices that use a physical barrier to separate cells, both viable and dead, from the cell culture medium. This allows high separation efficiencies. The major drawback of separation by size is that these systems tend to foul and clog. In the case of retention by density, the separation relies on the density difference between cells and culture medium without requiring filtering devices, thus making it more robust and suited for long-term operation. However, retention by density often comes along with long residence times of the cells within the retention device due to the rather small density difference between cells and medium, leading to low gravitational settling velocities (1–15 cm/h) (Voisard et al. 2003).

The time the cells spend in the retention device is, in fact, an important variable which should be kept rather short. The environment in these units is not controlled and differs from the one inside the bioreactor. Therefore, it might irreversibly impact cellular performance, as a drop in temperature, oxygen and nutrient limitations, or lower pH levels can affect cell metabolism.

On the whole, it is seen that a reliable cell retention device has to fulfil several requirements. These include ideally full cell retention, prevention of product sieving, operation at feasible perfusion rates, sufficient oxygen supply, prevention

ATF   TFF

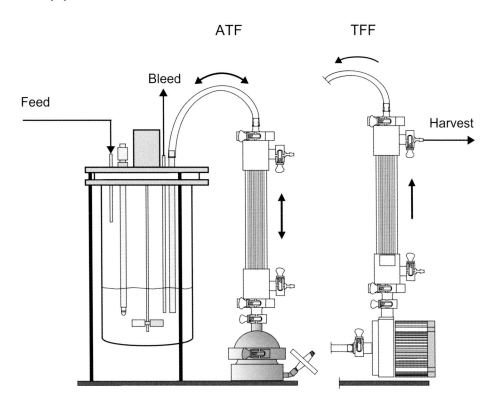

Figure 2.1 Schematic representation of the set-ups used for the operation of the ATF (left) and TFF (right) retention devices in continuous manufacturing. © Reprinted from *Biochemical Engineering*, 110, Daniel J. Karst, Elisa Serra, Thomas K. Villiger, Miroslav Soos and Massimo Morbidelli, 'Characterization and comparison of ATF and TFF in stirred bioreactors for continuous mammalian cell culture processes', 2016, with permission from Elsevier

of cell harmful shear stress, relatively short residence time and long-term operation without failure.

Today, tangential filtration devices such as the alternating tangential flow (ATF) and the tangential flow filtration (TFF) hollow-fibre systems represent industrial standard technologies and are typically preferred over the other ones. In the case of the ATF system, a diaphragm pump is connected to a hollow-fibre module (typical pore size: 0.22 μm) that is directly connected to the bioreactor system. The pump uses alternating back and forward pushes to exchange the culture broth between the reactor and the hollow-fibre module. In the case of the TFF system, a pump is directly connected to the bioreactor outlet and to the hollow fibre and moves the cell culture broth into an external loop, through the hollow fibre and back to the bioreactor system. The two operations are illustrated in Figure 2.1. In both systems, high liquid flowrates at the membrane surface can reduce filter fouling and keep the operation stable. In particular, it has been shown that the alternating flow within the ATF device leads to a specific beneficial self-cleaning effect of the fibre module (Bonham-Carter & Shevitz 2011). Furthermore, these technologies have advanced in recent years and nowadays commercial solutions exist for lab to manufacturing scale (1–1,000 L) consisting either of a single-use or a stainless steel device.

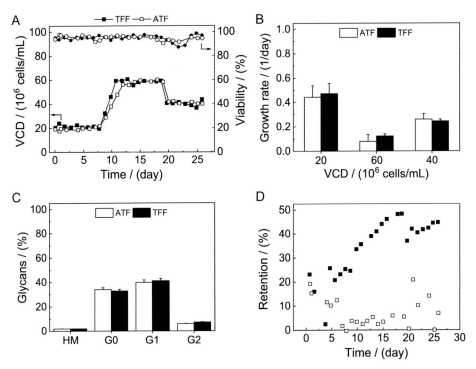

Figure 2.2 Comparative studies of the ATF and the TFF device on benchtop bioreactors using the same hollow fibre at three different viable cell density set points: (A) Viable cell density and viability, (B) Specific growth rates, (C) N-Glycan distribution and (D) product retention. ©Reprinted from *Biochemical Engineering*, 110, Daniel J. Karst, Elisa Serra, Thomas K. Villiger, Miroslav Soos and Massimo Morbidelli, 'Characterization and comparison of ATF and TFF in stirred bioreactors for continuous mammalian cell culture processes', 2016, with permission from Elsevier

Comparative studies of the ATF and the TFF devices using the same hollow-fibre have been reported by Karst et al. (2016). A laboratory bioreactor equipped with either the TFF or the ATF device was operated through three consecutive steady states, with VCD set points equal to 20, 60 and 40×10^6 cells/mL, respectively, as shown in Figure 2.2A. While process performance (cellular growth and metabolic input and output rates) as well as product quality of the produced mAb were similar between the two systems – as illustrated in Figure 2.2B and C, respectively – Figure 2.2D shows that a significantly larger portion of the protein was retained in the bioreactor equipped with the TFF device.

In a different study, Wang et al. (2017) also compared the process performance of the ATF and TFF devices. In the first step, they replaced a peristaltic pump by a low-shear centrifugal pump to operate the TFF, which dramatically reduced product sieving and led to comparable levels of sieving between the ATF and TFF systems. Karst et al. (2016) and Wang et al. (2017) used a similar type of centrifugal pump for their comparative studies but apparently obtained different results. While Karst et al. (2016) observed higher product retention in the TFF device, Wang et al. (2017) obtained similar levels of product retention in the ATF and TFF devices. This is, in fact, a system-dependent behaviour, so considering different cell lines, medium compositions and protein products may lead to different observations.

Using ATF and TFF retention devices in connection to wavebag bioreactors revealed that very high cell densities are limited by the vacuum capacity of the ATF, thus resulting in a maximum viable cell density of 1.32×10^8 cells/mL, compared to 2.14×10^8 cells/mL in the TFF device (Clincke et al. 2013a). This upper bound was imposed in the TFF by the high pressure inside the TFF loop due to the increased viscosity of the cell broth at such high flowrates. Additionally, too high pCO_2 levels and insufficient oxygenation were observed. On the other hand, comparing the ATF device to an internal spin filter system, Bosco et al. (2017) observed that the ATF system supports higher maximum perfusion rates and, consequently, higher maximum cell densities.

In acoustic filter systems, cell-free harvest is enabled by ultrasonic separation of cells and medium. The exposure to an acoustic resonance field leads to an increase in the biomass particulate size as a result of three different radiation forces inducing cellular aggregation. Separation efficiencies of greater than 90 per cent were achieved for cell densities below 10×10^6 cells/mL and high perfusion rates (Baptista et al. 2013, Gorenflo et al. 2003). However, this technique shows a narrow range of allowed operating conditions for optimal cell retention the cell separation efficiency is limited by cell concentration and input power, and process scale-up is difficult (Pörtner 2015).

This variety of studies shows the importance of cell retention in the development and design of perfusion bioreactors and highlights ATF and TFF systems as the current state-of-the-art technology at lab and manufacturing scale. Note that, just like sensors, the cell retention devices need to be resistant to sterilisation procedures (e.g., autoclaving and gamma sterilisation).

2.2.4 Configuration

Different lab-scale systems for the development of perfusion processes have been described in the literature. One example that we consider as being a good reference is the system developed by Karst et al. (2016). This consists of a stirred tank reactor (2.5 L glass vessel), typically operated at 1.5 L and equipped with a single six-blade Rushton turbine (RT) impeller that allows breakage and complete dispersion of gas bubbles at reasonable impeller speeds (200–400 rpm). Together with an open pipe sparger, this system allows both high oxygen mass transfer coefficients and high volumetric gas flowrates. The stirred tank reactor was coupled with either an ATF or a TFF cell retention device and then placed on a balance to implement weight-based flowrate control. A suitable head plate configuration in such a set-up includes the connection of the bioreactor to the cell retention device, all sensors and sufficient ports for the different inlet and outlet streams. The bioreactor set-up and head plate configuration are illustrated in Figure 2.3.

The developed system showed high mixing effectiveness with low shear and suitable gas–liquid mass transfers, with $k_L a$ values in the range 5–40 h^{-1}, sufficient to support up to 10^8 cells/mL (Ozturk 1996). A more detailed characterisation of reactor set-ups in terms of hydrodynamic stress, gas–liquid mass transfer, and mixing effectiveness, particularly at larger scales, is discussed in Chapter 5.

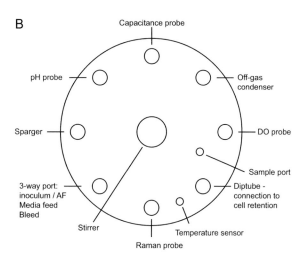

Figure 2.3 Overview of the bioreactor set-up (A) side view and (B) schematic of the head plate configuration.
© Reprinted/adapted by permission from: Springer Nature, Continuous and Integrated Expression and Purification of Recombinant Antibodies by Sebastian Vogg, Moritz K. Wolf and Massimo Morbidell (2018)

2.3 PERFUSION-SPECIFIC CONTROL VARIABLES: P, VCD AND CSPR

In this section, the most important parameters that characterise perfusion processes and that are necessary for process control are discussed. In these processes, the flowrates are commonly expressed as specific rates – that is, volume of medium per volume of bioreactor per day (VVD) or simply reactor volume (RV) per day. Consequently, the perfusion rate is the ratio of the volumetric feed flowrate, Q_{feed}, and the total reactor volume, V_R:

$$P = \frac{Q_{feed}}{V_R}. \tag{2.1}$$

In order to keep the reactor volume constant, the sum of the rate of cell-free supernatant removal – the harvest rate, H – and the rate of cell discard – the bleed rate, B – need to equal the perfusion rate:

$$P = H + B. \tag{2.2}$$

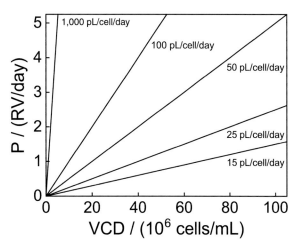

Figure 2.4 The CSPR describes the ratio of the perfusion rate (P) and the viable cell density (VCD). The lower the CSPR, the flatter is the slope of the straight line and, thus, the less medium is provided per cell.

A very significant operating parameter in perfusion processes is the cell-specific perfusion rate (CSPR), which relates the perfusion rate and the viable cell density (VCD), X_V and expresses the amount of medium fed to a single cell per day:

$$\mathrm{CSPR} = \frac{P}{X_V}. \tag{2.3}$$

The CSPR is generally expressed in pL/cell/day or nL/cell/day. In Figure 2.4, the relation between perfusion rate and viable cell density is illustrated. It is seen that the lower the value of the CSPR, the lower the perfusion rate is to maintain a certain viable cell density. For example, in the case we want to operate at a viable cell density of 20×10^6 cells/mL with a CSPR of 100 pL/cell/day, the perfusion rate has to be set at 2 RV/day. When we choose the CSPR to be 50 pL/cell/day for the same VCD, the perfusion rate has to be set to 1 RV/day. A further decrease to 25 pL/cell/day would require a decrease of the perfusion rate to 0.5 RV/day.

It is clear that in order to sustain a certain VCD value, it is necessary to provide sufficient nutrients. This implies that there is a minimum value of the CSPR, below which the operation becomes unstable and cells start to die. This is why the CSPR represents a key quantity in the design, optimisation and control of perfusion bioreactors, and it should be kept above such a minimum value, often referred to as minimum cell-specific perfusion rate (CSPR$_{\mathrm{min}}$). Accordingly, this provides a direct performance indicator for the given medium and expression system: the so-called medium depth. By maintaining such a CSPR value constant, cell metabolism and metabolite concentrations should remain constant, thus leading to a stable culture behaviour (Konstantinov et al. 2006, Ozturk 1996).

Reported minimum cell-specific perfusion rates range from 15 to 500 pL/cell/day (Dowd et al. 2003, Karst et al. 2016, Konstantinov et al. 2006, Warikoo et al. 2012, Xu & Chen 2016). Recent operations at very low CSPRs, about 15–20 pL/cell/day, indicate the improvement in medium formulations in recent years (Karst et al. 2017a, Xu & Chen 2016). The historical development of feasible CSPRs is illustrated in Figure 2.5.

It is worth pointing out that the previously mentioned operating variables need to respect very strict boundaries. The perfusion rate is lower bounded by the maximum allowable residence time in the reactor – and, therefore, by the protein stability – and

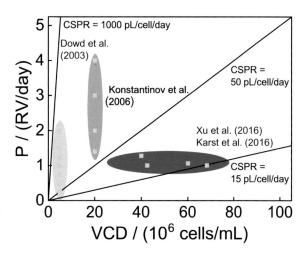

Figure 2.5 Historical development of CSPR operating ranges. In the early 2000s, Dowd et al. (2003) tested CSPRs in the range 50–400 pL/cell/day at 5×10^6 cells/mL (light blue), while Konstantinov et al. (2006) operated at CSPRs in the order of 50–400 pL/cell/day at 20×10^6 cells/mL (blue). In most recent studies, Karst et al. (2016) and Xu & Chen (2016) performed perfusion studies at CSPR values below 20 pL/cell/day (dark blue).

upper bounded by the volumetric capacity of the cell retention device. On the other hand, the maximum allowable VCD is directly linked to oxygen supply so that, in most cases, the maximum viable cell density (VCD_{max}) is limited by the maximum possible oxygen transfer rate (OTR) provided by the set-up. In addition, the overall biomass content in the bioreactor is upper bounded by the viscosity of the suspension, which affects mixing and all mass transfer processes.

2.4 CONTROL

High cell densities and elevated metabolic demands make continuous perfusion processes highly dynamic, and a robust process control is inevitably important. For a stable operation of the perfusion culture, the different flowrates (feed, harvest and bleed), the working volume and the cell density have to be tightly controlled. Special attention is given to the cell-specific perfusion rate independent of transient or steady-state operation, since this determines the stability of the culture. Altogether, several control strategies have to be combined – including classical control loops of pH, temperature, aeration and mixing – with perfusion-specific ones needed to keep inlet and outlet flows, as well as the liquid level in the reactor, constant.

2.4.1 Cell-Specific Perfusion Rate

As previously mentioned, the CSPR defines the amount of nutrient delivered to each cell per unit time. As such, it is a very important parameter for the culture operation. For example, at the (N-1) seed stage, high CSPR values are used to promote cellular growth at the expenses of an increased medium consumption. On the other hand, at the N production stage, a lower CSPR reduces nutrient delivery to the cells, leading to not only decreased cellular growth rates but also higher titres and more efficient medium utilisation.

The proper design and development of a mammalian cell perfusion process requires a careful analysis of suitable viable cell densities, perfusion rates and cell-specific perfusion rates for a given set-up, expression system and medium. Indeed, maintaining the desired CSPR during operation is needed to have a stable and efficient process. Nevertheless, this does not mean that we need to implement a control loop having CSPR as the set point. In the following section, we describe control strategies based on the control of P and VCD, with set points which obviously satisfy the desired CSPR value (Konstantinov et al. 2006), although alternative strategies are also possible (Konstantinov et al. 1996, Ozturk 1996).

2.4.2 Primary Control Loops

The bioreactor, independently of the batch or continuous operating mode, has to be operated at a fixed set of operating conditions including DO concentration, pH, temperature and stirring. Consequently, temperature control; mass flow control for air, nitrogen and oxygen to maintain the desired DO; and a tight control of the stirring speed have to be installed since the failure of any of these might have an irreversible impact on the culture. In general, similar systems to those currently used for fed-batch bioreactors can serve this purpose also in the case of perfusion cultures.

Typically, cell cultures should be operated at pH 6.9–7.2 (Miller et al. 1988, Wong et al. 1992). The control of the cultures pH is often implemented using a buffer system with CO_2 addition to decrease pH and base addition to increase pH. However, it has been demonstrated that the understanding of the interactions between the partial carbon dioxide pressure, pCO_2; lactate, a by-product of the cells; and pH can enable the operation of stirred-tank bioreactors without external pH control, thus preventing the addition of excessive amounts of base (Xu & Chen 2016). A different approach is to use only a one-sided pH control strategy by controlling the CO_2 fraction in the inlet gas, thus again preventing excessive base addition (Karst et al. 2016).

The dissolved oxygen is controlled by the fraction of oxygen (oxygen partial pressure, pO_2) and the volumetric flowrate of the inlet gaseous stream. However, too-high oxygen partial pressures can stimulate the generation of radical oxydative species (ROS) and induce cell death. Consequently, sparging of pure oxygen should be prevented (Godoy Silva et al. 2010). Commonly used process set points include a DO of 50 per cent, temperature at 36–37°C and stirring rates in the range 200–400 rpm.

2.4.3 Control of Working Volume and Perfusion Rate

The stability and robustness of perfusion cultures inevitably depends on the tight and robust control of the working volume and the perfusion rate. These two parameters are directly related. For working volume control, either the reactor has to be placed on a balance or a sensor that monitors the liquid level inside the reactor has to be used. Two control strategies can be adopted: the perfusion rate is fixed, and the harvest rate is adjusted by a proper controller so as to keep the measured weight or liquid level in the reactor constant (Figure 2.6) or, vice versa, the harvest rate is maintained constant while the feed rate is properly adjusted by the controller (Figure 2.7). Adjustment of either the perfusion/feed or the harvest rate according to the reactor weight can be regulated by a standard proportional integral derivative (PID) controller.

The choice of one or the other strategy depends on the user's application and priorities. During the process-development phase, it could be recommended to keep constant the perfusion rate. Once the process is fixed, all expected flowrates are known, and this issue becomes less important. On the other hand, in the case where the bioreactor is directly coupled to the first step of the downstream purification train – e.g., the capture step, as in the context of continuous integrated

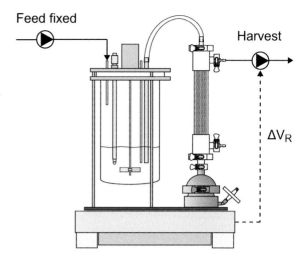

Figure 2.6 Control of working volume and harvest rate: operation at fixed perfusion/feed rate with proper adjustment of the harvest rate to control the reactor weight. © Reprinted from *Biochemical Engineering*, 110, Daniel J. Karst, Elisa Serra, Thomas K. Villiger, Miroslav Soos and Massimo Morbidelli, 'Characterization and comparison of ATF and TFF in stirred bioreactors for continuous mammalian cell culture processes', 2016, with permission from Elsevier

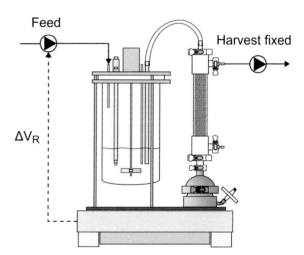

Figure 2.7 Control of working volume and feed/perfusion rate: operation at fixed harvest rate with proper adjustment of the feed/perfusion rate to control the reactor weight. © Reprinted from *Biochemical Engineering*, 110, Daniel J. Karst, Elisa Serra, Thomas K. Villiger, Miroslav Soos and Massimo Morbidelli, 'Characterization and comparison of ATF and TFF in stirred bioreactors for continuous mammalian cell culture processes', 2016, with permission from Elsevier

manufacturing – it is probably better to keep constant the harvest rate, which constitutes the feed flowrate to the capture unit so as to avoid propagating possible disturbances along the downstream units.

The cell containing bleed stream is necessary to compensate for cell growth and achieve steady state. Since the bleed stream contains also the target protein, which is wasted, it is directly linked to product loss and should be kept as small as possible to maximise the process yield. The bleed rate is usually not directly included in the weight control strategy of the bioreactor as it constitutes 10 to 20 per cent of the overall perfusion rate (Deschênes et al. 2006, Lin et al. 2017). Its control is best implemented through a biomass control strategy, as described in the next section.

2.4.4 VCD/Biomass Control

The viable cell density in perfusion bioreactors is typically controlled by the bleed rate. For this, various methods for monitoring the viable cell density are used – ranging from acoustic resonance densitometry, conductivity, optical sensors and

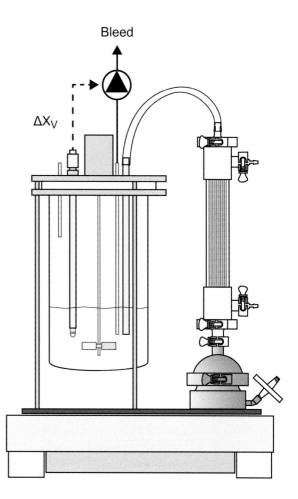

Figure 2.8 VCD/Biomass control strategy through the bleed rate. © Reprinted from *Biochemical Engineering*, 110, Daniel J. Karst, Elisa Serra, Thomas K. Villiger, Miroslav Soos and Massimo Morbidelli, 'Characterization and comparison of ATF and TFF in stirred bioreactors for continuous mammalian cell culture processes', 2016, with permission from Elsevier

real-time imaging to indirect methods such as the measurement of the oxygen uptake rate or the adenosine triphosphate (ATP) production rate (Carvell & Dowd 2006, Deshpande & Heinzle 2004, Dowd et al. 2003, Kiviharju et al. 2008, Konstantinov et al. 1994, Yoon & Konstantinov 1994). The current state-of-the-art technology is based on capacitance sensors for the real-time monitoring of viable cell density.

In order to operate at a constant viable cell density, two approaches are possible. The simpler one involves a semi-continuous bleeding strategy, which is applied when no online measurement of the cell density is available. In this case, a proper portion of the bioreactor cell broth is removed once or twice per day according to measured viable cell densities, as discussed by Clincke et al. (2013a) and Karst et al. (2016). For the second approach, online monitoring of the biomass or viable cell density is necessary. The signal of the biomass probe can be used to implement an automated bleeding strategy (Deschênes et al. 2006, Konstantinov et al. 1994, Steinebach et al. 2017) as schematically indicated in Figure 2.8, where the bleed rate is adjusted through a standard PID controller to keep the biomass constant at the set point value.

It is worth noting that the system responds differently to changes in the bleed rate, depending on whether the perfusion or the harvest rate is kept constant – i.e., Figures 2.6 and 2.7, respectively. For example, in the first case, an increase in the bleed rate leads to a decreased harvest rate. On the other hand, if the harvest rate is set

Figure 2.9 Overall bioreactor control strategy. The harvest rate is fixed and the bleed pump is controlled through a proper VCD sensor. The feed pump is adjusted to keep constant the reactor weight. © Reprinted from *Biochemical Engineering*, 110, Daniel J. Karst, Elisa Serra, Thomas K. Villiger, Miroslav Soos and Massimo Morbidelli, 'Characterization and comparison of ATF and TFF in stirred bioreactors for continuous mammalian cell culture processes', 2016, with permission from Elsevier

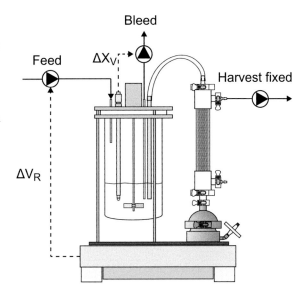

constant, as the bleed increases, the perfusion/feed rate also increases, as illustrated in Figure 2.9.

2.4.5 Cell Retention Device Operation

The reliable operation of the cell retention device is a strict requirement for long-term stable perfusion cultures. For this, close monitoring and control of the exchange rate between the bioreactor and the external retention device is necessary so as to provide early indication of operational failure. In the case of the alternating tangential flow module, a diaphragm pump typically is used. This pump is controlled by a separate controller supplied by the manufacturer. The periodic alternation of pressurising and depressurising cycles results in an oscillating change of the flow direction and rate. On the other hand, in the TFF device, where the cell culture broth is pushed by a centrifugal pump through an external loop containing the retention device, it is important to carefully control the rotational speed of the pump. High rotational speed can lead to shear stresses that exceed the cells maximum tolerable stress (Karst et al. 2016, Neunstoecklin et al. 2015).

An important aspect refers to the monitoring of the degree of fouling of the hollow-fibre membrane. For this, additional pressure sensors can be mounted on the permeate and the retentate side of the membrane to measure the transmembrane pressure. An increasing transmembrane pressure and decreasing permeate flow typically indicate progressive filter fouling (Kelly et al. 2014). At lab scale (1–10 L bioreactors), exchange rates of 1.0–1.5 L/min are typically reported (Clincke et al. 2013a, Karst et al. 2016).

2.5 MEDIA FOR PERFUSION CELL CULTURES

Medium formulation plays an important role in the development and optimisation of any cell culture process. In the early days of cell culture, media consisted of natural biological substances like plasma, serum and other animal or plant extracts, which naturally contain the nutrients and growth factors necessary for the cells to grow. The use of these ingredients, however, posed several problems, including the lack of control over their composition, resulting in significant lot-to-lot variations and, consequently, unreliable cell culture processes. In addition, due to their animal origin, these components could carry pathogens like viruses, thus making these media too risky for the production of active species that would eventually be injected into a patient (Grillberger et al. 2009, Yao & Asayama 2017).

Modern cell culture media are mostly chemically defined (CD); that is, all the components are well-defined chemical species dosed at a precise concentration (Yao & Asayama 2017). This includes, for example, salts, vitamins and amino acids but also, in certain cases, recombinant proteins (highly purified and produced in a controlled environment). CD media include up to hundreds of different ingredients, and their formulation for a given culture involves the definition of all the relative concentrations. This is indeed a complex task which requires specific knowledge on cell biology (Lin et al. 2017, Ling 2015, Ozturk & Hu 2006, Reinhart et al. 2015, Yao & Asayama 2017).

The existing knowledge about media design and optimisation for fed-batch processes can be leveraged for the use in perfusion processes (Lin et al. 2017). Fed-batch media are optimised neither for the cellular needs in a perfusion process nor for some operational constraints related to the large volumes that need to be prepared, sterilised and stored before being added to the bioreactor. This is a particularly relevant point, as a proper medium design has a significant impact on process operation and constitutes a significant, if not the largest, portion of the process operating cost (Pollock et al. 2013, Xu et al. 2017b).

In the following, we discuss a few aspects specifically related to media application in perfusion, with reference to the medium depth and the medium concentration, which are both related to maximising the reactor volumetric productivity by using more efficient media. The medium depth defines the minimum perfusion rate of that medium needed to sustain a given viable cell density and corresponds to the reciprocal of the medium-specific $CSPR_{min}$ value (Konstantinov et al. 2006):

$$\text{Medium depth} = \frac{1}{CSPR_{min}} \qquad (2.4)$$

For a given medium formulation – that is, for a given medium depth or $CSPR_{min}$ – the sustainable VCD in a perfusion bioreactor can be increased by operating at higher perfusion rates in order to provide the same amount of nutrients per cell. This corresponds to moving along the iso-CSPR lines shown in Figure 2.4, with the one corresponding to the $CSPR_{min}$ obviously providing the most convenient condition – that is, the largest possible viable cell density for the given perfusion rate. Improving the medium depth targets to decrease the $CSPR_{min}$ – and, therefore, to increase the viable cell density for a given perfusion rate – which leads to a higher volumetric productivity of the reactor.

Chemically defined media offer the possibility of taking a systematic approach to media design by investigating the effect of each individual species on the medium depth. The various ingredients can be grouped into different categories that include carbon sources, amino acids, vitamins, trace elements, salts, lipids, polyamines and hormones. Some of these are consumed by the cells as an energy source or as building blocks for housekeeping activities, cellular division and, of course, for the synthesis of the targeted recombinant protein. Other functions do not imply the component depletion, like shear stress protectants or salts containing sodium or potassium,

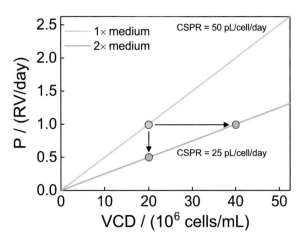

Figure 2.10 Relation of CSPR, P and VCD for a given cell line and two different media concentrations: one (2× medium) being the double of the other (1× medium).

which act on osmolality and allow the cells to build up the natural sodium potassium transmembrane gradient. Since each component can be quantified individually, it is possible to monitor its consumption during the cell culture and consequently rebalance the medium by adding some concentrated feed formulations containing all or parts of the media components. This allows grouping media components into two main categories:

• Nutrients: amino acids, vitamins, carbon sources
• Baseline components: pluronic, salts, buffers

Indeed, the selection of the hundreds of species present in a medium and of their relative composition is a difficult task since, in principle, each of them can affect the medium depth. However, since we defined the CSPR as the volume of media per cell per unit time, the medium concentration also affects the CSPR. That is, by concentrating all nutrients in the medium, we can feed to the reactor the same mass of each nutrient but with a lower volumetric perfusion flowrate, thus effectively reducing the $CSPR_{min}$ or increasing the medium depth. Medium concentration is, therefore, also an important parameter in medium design. For example, assuming that there are no physical (i.e., solubility) limitations in doubling the concentration of all medium compounds, going from a 1× medium to a 2× medium, the CSPR decreases by a factor of 2, e.g., from 50 to 25 pL/cell/day, as shown in Figure 2.10. As a result, the same cell density can be sustained by half the perfusion rate, or, alternatively, the VCD can be doubled while keeping the perfusion rate unchanged – all of this, of course, under the assumption that physiological conditions are maintained and that no other limitation, such as oxygen consumption or accumulation of toxic metabolites, arises.

However, in practice, increasing media concentration is not straightforward as several limitations must be accounted for, with solubility and stability being the primary concern. Physiological conditions imposed by the biology of the cells also include pH and osmolality constraints that the bioreactor environment cannot violate. In general, this is achieved by buffering the pH using bicarbonate and adjusting the osmolality using salts. As an example, in Figure 2.11, it is shown how to double all components in the medium (glucose, vitamins, amino acids and others), while adapting sodium chloride to keep osmolality unchanged. This

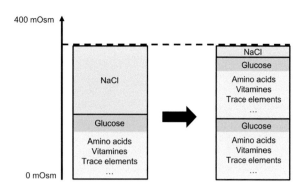

Figure 2.11 Schematic of media composition expressed in osmolality contribution: the concentration of sodium chloride is adapted to maintain osmolality unchanged while doubling the nutrient concentrations.

example is simplified and does not consider, for example, the adaptation of the pH buffer that would, in practice, change because of the increased proportion of acidic compounds.

Thus, when designing a medium for a given culture, we need to consider the function of each of its components. Metabolic needs can be understood by comparing the composition of the spent medium in the harvest stream to that of the fresh medium. From this, cell-specific consumption and production rates of the different nutrients can be computed to estimate the relative need of each nutrient. The chemical species that buffer pH, control osmolality or act as shear protectant do not necessarily need to be concentrated in order to improve the medium depth. Nevertheless, their concentration should also be adapted and optimised in the medium according to their function. For example, if the amino acids, glucose and vitamins are concentrated, the salt concentrations should be reduced to keep the osmolality constant. On the other hand, the bicarbonate level should be adapted to the new formulation to maintain its buffering action unaltered. The concentration of the shear protectant should be maintained roughly constant, usually between 0.5 and 2 g/L, although higher concentrations up to 5 g/L have been reported to be beneficial, especially at higher cell densities (Meier et al. 1999, Xu et al. 2017c). Of course, these considerations are of very general value but provide the basis for fine-tuning the concentrations of the different ingredients in the medium and improving the volumetric productivity or the footprint of a given cell culture.

In an industrial manufacturing environment, media are prepared starting from a homogeneous powder formulation, which typically represents the company's platform medium and contains most, if not all, of the compounds. This leaves no room for flexibility, for example, to concentrate the nutrients without simultaneously concentrating also the baseline components. On the other hand, in a process development set-up, it is more convenient not to include baseline components in the powder formulation to leave open the possibility of concentrating the nutrients while independently changing or keeping constant the baseline components. By adding a certain portion of such a depleted powder to a defined volume of liquid, a $1\times$ concentrated medium is obtained. To concentrate it twice, two parts of the same powder should be added. Baseline compounds are not part of this powder formulation, and therefore, their concentration can be adjusted independently, depending of their specific function, as schematically illustrated in Figure 2.12. This implies controlling osmolality with the salts, pH with the buffer and other characteristics of the culture with other components. This requires careful dosage of the powder (nutrients) and of all the other components to obtain an efficient

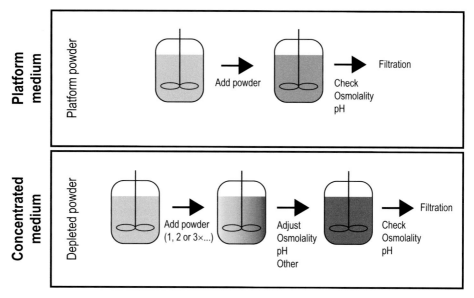

Figure 2.12 Example of media preparation. Production environment (top): the platform powder containing all medium compounds is solubilised, and after quality check (pH and osmolality), the medium is filtered and used. Process development environment (bottom): the depleted powder is added to the extent needed for a certain nutrient concentration, and then the formulation is adjusted (pH and osmolality) using baseline components before being filtered and used.

medium for the overall needs of the cell culture. In practice, this is typically done by properly combining a so-called basal medium, which corresponds to a formulation that is concentrated $1\times$ and includes all the baseline compounds, with formulations containing only nutrients with higher concentrations, as discussed in the following example.

Let us consider, for example, the case where we are ideally operating the bioreactor at the $CSPR_{min}$ and at a certain VCD value, and we now want to increase the VCD twice. As discussed earlier, we can double the perfusion rate so as to keep constant the CSPR value and not drop it below its minimum value, which would lead to an unstable culture. An alternative option would be to concentrate the medium formulation to decrease the $CSPR_{min}$ of the given medium and cell line combination. A third option that would allow using a fixed basal medium would be the use of a secondary feed, highly concentrated in only the nutrients. The flowrate of this feed would be negligible or at least very low compared to that of the basal medium, but the overall medium depth, when considering the total perfusion rate given by the basal medium and the secondary feed, could be significantly increased. An example of such an application was given by Karst et al. (2016), although in that case, the concentrated feed was premixed to the basal medium. This approach provides a very flexible basis to change the medium depth while using well-defined and fixed streams: the basal medium and a concentrated nutrient secondary feed. Another advantage is that solubility or stability limitations faced by some components in the complex medium formulation can be reduced in the secondary feed. Solubility of amino acids, for example, varies with pH and may become limiting for the medium stability (McCoy et al. 2015). The pH of the feed can, in fact, be adjusted according

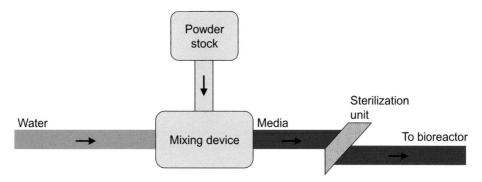

Figure 2.13 Schematic of an online media preparation device.

to solubility limitations without being critical since it is used in combination with the medium that has some pH buffer capacity.

Finally, it is worth to at least mention the possibility of preparing media by sterile online mixing – for example, from a single powder, as shown schematically in Figure 2.13. Using such a device, media could be produced on-demand and storage space would be reduced significantly. From a technical point of view, such a device has to fulfil several requirements, such as tight control of the homogeneity of the solution and of the added amount of powder. In addition, a careful characterisation of mixing times, pH adjustments and other media-specific constraints (solubility and buffer capacity) are necessary and potentially challenging to realise.

2.6 STEADY-STATE OPERATION AND PROCESS DYNAMICS

2.6.1 The Continuous Stirred Tank Reactor Model

The aforementioned equipment and control strategies enable the continuous operation of mammalian cell perfusion cultures at constant viable cell density and provide the basis for steady-state operation. Assuming uniform composition and temperature within the bioreactor, its behaviour can be described in the frame of the classical continuous stirred tank reactor (CSTR) model. With reference to a generic species i, the mass balance in the CSTR is given by the following equation:

$$V_R \frac{dt}{dc_i} = Qc_i^{in} - Qc_i + r_i V_R, \tag{2.5}$$

where V_R is the reactor working volume, assumed to be constant, Q is the volumetric flowrate; c_i is the concentration of the species i inside the reactor; $c_{i,in}$ is its concentration in the inlet stream; and r_i is the corresponding reaction production term. Dividing both terms by V_R and using Equations (2.1) and (2.2) allows to reformulate this equation in terms of the volume specific rates as follows:

$$\frac{dc_i}{dt} = Pc_i^{in} - (H + B)c_i + r_i, \tag{2.6}$$

where the term on the left-hand side represents accumulation inside the reactor, while the three terms on the right-hand side represent the contribution of inlet (perfusion), outlet (harvest and bleed) and production rate, respectively. A similar mass balance can be written for the viable cell density leading to

$$\frac{dX_v}{dt} = (\mu - B)X_v, \tag{2.7}$$

where the accumulation term is set equal to the difference between the cell growth rate, μ, and the bleed rate. The first one describes the cell-doubling process and is a function of the cell metabolism and consequently of the entire cell environment inside the reactor. Dividing Equations (2.6) and (2.7) by the viable cell density X_v allows calculating cell-specific accumulations:

$$\frac{dc_i}{dt}\frac{1}{X_v} = \frac{(Pc_i^{in} - (H+B)c_i)}{X_v} + q_i \tag{2.8}$$

$$\frac{dX_v}{dt}\frac{1}{X_v} = \mu - B, \tag{2.9}$$

where the cell-specific production rate of the species i, q_i, represents the ratio r_i/X_v and is related not only to chemical reactions but also to biological processes involving the cell metabolism.

By operating a CSTR with constant inlet flows, it is possible, after a certain transient period of time, to achieve a stationary state, where the inlet streams and production processes balance the outlet streams and consumption ones. In this state, all the process variables are invariant in time and the environment inside the reactor, which was already uniform in space, also becomes constant in time. This is the ideal situation for the cells to always follow the same path in expressing the target protein. Therefore, this leads to a product quality heterogeneity, in terms of glycoform and charge variant distributions, which is reduced to the minimum compatible with the living nature of the cells. In contrast, in a fed-batch process, the environment changes during the entire process, with depletion of the nutrients and accumulation of the toxic metabolites. This affects the cell metabolism, leading to differences in the target protein expressed during the run, and then increasing heterogeneity of the corresponding quality attributes. As a consequence, product quality attributes are much more heterogeneous in batch than in continuous operations.

At steady state, there is no accumulation of any species inside the reactor, and consequently, mass balances 2.8 and 2.9 reduce to

$$q_i = \frac{(H+B)c_i - Pc_i^{in}}{X_v} \tag{2.10}$$

$$\mu = B. \tag{2.11}$$

It is worth pointing out that the notion of steady state in living systems has to be taken with some care. In particular, the cell-specific productivity, q_i, and the cell growth rate, μ, are functions of the cell metabolism and therefore may be affected by several processes related to cell ageing or high cell generation numbers. In some cases, for example, this has been observed to lead to the decrease of the cell-specific productivity of a mAb over the culture time (Karst et al. 2017a, Ozturk 2014). In general, these processes are, however, rather slow and occur over several days. In particular, they are much slower than the bioreactor characteristic time, which is related to the average residence time and is in the order of one day. This implies that the bioreactor dynamics is usually fast enough to adapt to the slow changes of q_i and μ, thus establishing so-called pseudo-steady-state conditions. This is an important general concept in system dynamics, which means that the accumulation terms, although non-zero, are negligible with respect to the other terms in the mass balances, so Equations (2.10) and (2.11) still hold true, with all the involved terms changing slightly in time. In other words, the input and output terms in the mass balances compensate each other instantaneously (the reactor dynamics) although they all change on a long time scale (the cell metabolism dynamics).

2.6.2 Residence Time Distribution

In a perfusion bioreactor, fresh and spent medium are continuously exchanged. The time that every single molecule spends inside the bioreactor system is described by the residence time distribution (RTD). The average residence time, τ, is given by the ratio between the reactor volume, V_R, and the volumetric exchange rate, Q, which is equal to the inverse of the normalised perfusion rate:

$$\tau = \frac{V_R}{Q} = \frac{1}{P}. \tag{2.12}$$

As mentioned earlier, a perfusion bioreactor can be described through the CSTR model, and therefore, the corresponding residence time distribution is given by the typical exponentially decreasing function:

$$\text{RTD} = E(t) = \frac{1}{\tau} \times exp^{-\frac{t}{\tau}}. \tag{2.13}$$

The preceding equations allow quantifying the influence of the perfusion rate on the residence time of all molecules inside the reactor. The higher the perfusion rate, the shorter the average residence time value and the sharper its distribution. For example, for medium perfusion rates going from 0.5 to 2 RV/day, the fraction of molecules that stays in the reactor less than one day increases from 39 to 86 per cent, as shown in Figure 2.14A and B. By comparison, in a classical fed-batch process lasting 14 days, the RTD looks quite different, as shown in Figure 2.14C. The protein is produced over the entire run, and, consequently, each single molecule has a different exposure, ranging from zero to 14 days, to post-translational modifications. This leads to very heterogeneous product quality attributes, since proteins produced in early stages are more likely to be modified than proteins produced at later stages. This, in addition to the effect of the changing cell environment discussed earlier, further contributes to the fact that product quality attributes are much more heterogeneous in batch cultures than in continuous cultures.

2.6.3 Steady-State Characterisation

Due to the presence of living cells, the processes taking place in a bioreactor are far more complex than those occurring in a typical chemical reactor. This also reflects on the definition of steady state and its characterisation. In particular, we need to distinguish between extracellular (e.g., glucose, lactate and amino acids) and intracellular processes which involve different chemical and biological entities and, therefore, have to be analysed separately. The chemical composition of the supernatant can be easily analysed, and it has been shown to become constant in time once steady state is achieved after a proper transient time, resulting in constant cell-specific production and consumption rates of the involved metabolites (Karst et al. 2017a). A detailed analysis based on metabolomics (Karst et al. 2017c, 2017d) has demonstrated that the concentrations of intracellular metabolites – such as nucleotides, nucleotide sugars and lipid precursors – also reach constant levels in a perfusion culture at constant cell density and perfusion rate. By analysing the cellular transcriptome and proteome, it has been confirmed that most of the identified transcripts and intracellular proteins fulfil the steady-state conditions (Bertrand et al. 2019). For a minority of transcripts, no steady state was achieved. Although no correspondence with a specific biological path could be identified, this may indicate that even if steady-state conditions can be achieved in perfusion operation, both intra- and extracellular, some biological events that occur with minor macroscopic effects may still be varying and changing over the time course of the operation.

Due to the complexity of these processes and the intrinsic variability of living systems, the operation of mammalian cell perfusion cultures at constant cell density

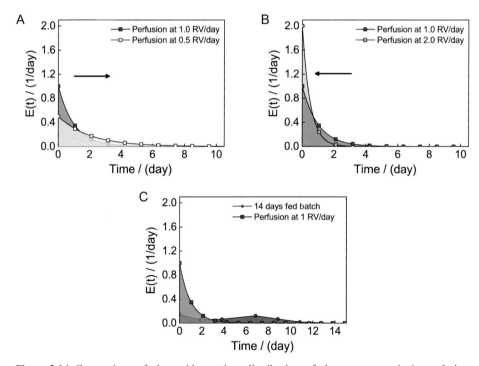

Figure 2.14 Comparison of the residence time distribution of the target protein in perfusion bioreactors at different perfusion rates and in a fed-batch bioreactor: (A) Decreasing the perfusion rate from 1.0 to 0.5 RV/day in a perfusion bioreactor leads to a broader RTD, (B) increasing the perfusion rate from 1.0 to 2.0 RV/day sharpens the RTD, and (C) in a 14-day fed-batch the RTD is very wide with an average value of 7 days, while more than 60 per cent of the produced proteins spend less than one day in a perfusion system operated at a perfusion rate of 1 RV/day.

and perfusion rate is sometimes referred to as state of control instead of steady-state operation (Nasr et al. 2017). This means critical process parameters and product quality attributes are kept in a well-defined narrow range of values with only small perturbations, rather than at a fixed value. Appropriate control strategies enable operation at this state, which allows the production of the target protein with sufficiently uniform quality.

2.6.4 Transient to Steady State

In a perfusion bioreactor, various processes characterised by different dynamics take place simultaneously, and their superimposition determines the dynamic behaviour of the whole unit. These include, on the one hand, the bioreactor hydrodynamics that is mainly defined by the medium exchange rate and, on the other hand, the cellular dynamics (affecting growth, consumption and production rates) defined by the cell metabolism.

The hydrodynamics of the reactor is expected to reach steady state faster, as the cellular response in a constant environment might lag behind. The reference point day zero is defined as the time where the VCD set point is reached. For a perfusion rate of 1 RV/day, it has been observed that extracellular metabolites take up to three days in order to achieve constant levels. This corresponds to three times the average residence time, τ, which is known to be the typical time for reaching steady state in a continuous stirred vessel. Similar trends have been observed for cell-specific rates as the growth rate and glucose uptake or lactate consumption rates, underlining that the cellular exchange rates with the environment follow the reactor hydrodynamics (Karst et al. 2016). With respect to intracellular CHO cell metabolites, Karst et al. (2017c) observed similar trends for the energetic state of the cell, as the energy charge, the uridine (U) and the nucleotide triphosphate (NTP) ratio achieved constant levels about three days after a set point change.

The detailed analysis of the intracellular proteome and transcriptome has revealed three major dynamics in a perfusion bioreactor operated at 1 RV/day, each corresponding to a different set of chemical and biological entities: one is completed and reaches steady state after about three days, another one after seven days and the last one never reaches a time constant state or, if it does, only after extremely long times. The first two are illustrated in Figure 2.15, taking as day zero the one where constant VCD is achieved. It is seen that the hydrodynamic transient of the reactor, and the corresponding extra- and intracellular metabolites mentioned earlier, is over in about 3τ. On the other hand, some of the product quality attributes exhibit a longer dynamics. In particular, patterns of the N-linked glycans took between five and seven days to reach stable levels, while charge isoform patterns took three to four days (Karst et al. 2017a, 2018; Walther et al. 2018). These results indicate that the reactor hydrodynamics dominate the dynamics of most of the process variables. However, some intracellular processes, like those involved in the synthesis of glycoforms, lag behind and take longer.

The longest observed dynamics is related to the intrinsic nature of living systems. As mentioned earlier in the context of Equations (2.10) and (2.11), instabilities of the cell lines due to genetic or epigenetic changes, while going through multiple generations or doublings, might lead to a decrease in cell-specific productivity. For example, Karst et al. (2016) observed a 30 per cent drop in productivity over 90 generations,

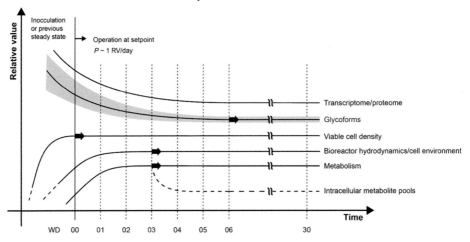

Figure 2.15 Different dynamics in a perfusion bioreactor reaching steady-state conditions. Constant viable cell density defines the reference day. Metabolism and bioreactor hydrodynamics take about three average residence times – i.e., three to four days, while product quality takes about six to seven days to reach steady state.

and even larger changes are reported by Ozturk (2014) during the entire process run, without never achieving steady-state conditions. However, these are typically very slow dynamics which, per se, do not cause significant difficulties with respect to the reactor operation, which can then be regarded as a pseudo-steady-state, as discussed earlier. On the other hand, in the long term, these processes reduce the bioreactor productivity and possibly also the product quality and, therefore, cannot be tolerated. This highlights the importance of developing robust and stable cell lines specifically dedicated to perfusion processes.

2.6.5 Product Quality in Batch and Continuous Reactors

Whether we refer to steady state or state of control, mammalian cell perfusion cultures clearly benefit from the reactor environment which is uniform in space and constant in time, as well as from the short and narrow residence time distribution of the target product. This results in a more uniform cell environment which promotes low product heterogeneity and shorter times for post-translational modifications. In addition, the continuous exchange of medium prevents the accumulation of inhibitory or toxic by-products, like lactate or ammonium. In contrast, in a fed-batch culture, the environment changes during the entire reactor run so that the cells experience a continuously changing environment, thus promoting corresponding changes in the product quality attributes, such as the patterns of N-linked glycosylation and charge variants. In addition, the broad RTD of the target protein contributes to this heterogeneity, leading to different modification degrees, depending on the time spent by the single protein inside the reactor after its expression. With illustrative purposes, in Figure 2.16, the charge variant distribution obtained in the same system but operated in the fed-batch and in the continuous mode are compared. It is seen that – as expected, based on the previously described processes – the distribution obtained in the continuous mode is narrower, thus indicating a less heterogeneous target protein (Karst et al. 2018).

Figure 2.16 Comparison of the chromatograms of charge isoform distribution for a commercial mAb produced in fed-batch and in perfusion mode. © Reprinted from *Current Opinion in Biotechnology*, 53, Daniel J. Karst, Fabian Steinebach and Massimo Morbidelli, 'Continuous integrated manufacturing of therapeutic proteins', 2018, with permission from Elsevier

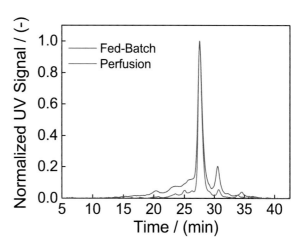

2.7 CONCLUSION

In this chapter, we have discussed the fundamentals of operation and control of perfusion bioreactors. Although these concepts are valid in general across scales, our reference was the bench-scale bioreactor at the litre scale. We have introduced the different equipment needed to realise a perfusion process and have indicated the critical aspects in the operation of each one of them. The most important operating and performance parameters have been considered, with particular reference to the CSPR, which is peculiar of the perfusion operation. Another important aspect in these systems is the media design, which needs not only to guarantee a sufficient supply of nutrients but also to keep physiological conditions during the entire operation – in particular, in terms of osmolality and pH. Although the currently used media are not specifically designed for perfusion operation, this can achieve large productivities based on high cell density and elimination of waste metabolites.

Perfusion bioreactors exhibit a distinctly different behaviour compared to batch or fed-batch units. This, in general, includes an initial transient behaviour followed by steady-state conditions which are ideal to produce consistently high quality and homogeneous products. In addition, continuous operation allows the achievement of very small residence times of the product inside the reactor, which limit all the following processes that affect the product quality, such as decomposition reactions or aggregation processes.

3

Scale-Down Models and Sensors
for Process Development

The design and development of perfusion cultures require extensive experimental campaigns in order to identify the most convenient conditions in terms of a very large number of operating parameters. This effort can be alleviated through the use of mathematical models, which are discussed in Chapter 6. Nevertheless, there is a need for high-throughput technologies that allow for the simultaneous operation and monitoring of several cultures at different process conditions. This requires suitable scale-down models, which allow the reliable prediction of the behaviour at larger scale. Within this chapter, we describe systems that allow the operation at the μL and mL scale and that enable early process development, such as clone and media screening, as well as the design of suitable process operating conditions at a larger scale. Managing such a large number of experiments is not possible without the proper level of automation, including suitable sensors for process monitoring.

It is to be said that these technologies, in the frame of perfusion process development, are still very young, certainly far from the level of maturity and therefore reliability of the corresponding scale-down models used for batch or fed-batch bioreactors. Accordingly, sometimes in this chapter, we discuss what would be desirable to have more than what is actually available.

3.1 PERFUSION CELL CULTURE AT DIFFERENT SCALES

Scale-down models are experimental devices used at the laboratory scale to reproduce the behaviour of larger units. In cell culture, the manufacturing scale ranges from tens to thousands of litres, whereas the scale-down models range from nanolitres to litres. The use of such devices is quite common in the biopharmaceutical industry to gather information needed in various moments of the development and operation of an industrial cell culture. This requires the screening of a large number of process conditions, such as for clones or media selection, investigation of various operating conditions, such as in process optimisation or process characterisation – e.g., during validation procedures.

The typical approach in the design of chemical and biotechnology processes is to first develop them in small units and then scale them up to a larger scale, which in biopharmaceutical industry means clinical and eventually commercial applications. This is done using either mathematical models or laboratory or pilot scale units or a combination of both, depending upon their convenience and accuracy. With respect to perfusion processes, the first step of this process is the identification of suitable process conditions enabling a stable culture, with an appropriate host cell line and media able to express the desired protein. For this, we use scale-down models (Long et al. 2014), whose results can then be reproduced at a larger scale. These devices, also often referred to as high-throughput units, derive their name from their ability to reproduce the behaviour of larger units.

In general, there are several requirements that need to be fulfilled. On the one hand, the model must be accurate enough to reproduce the larger scale fermentation process. On the other hand, it should offer a sufficiently high throughput and require a reasonably low workload. Mammalian cell cultures typically take days or weeks to be completed, and therefore, the model must be reliable and robust to minimise failures over long-term operation. In addition, cultures must be kept in a sterile environment to avoid any contamination by foreign and undesired microorganisms.

Like in any cell culture device, in scale-down models, the environment needs to be similar to physiological conditions. For example, in bench-scale bioreactors, there is typically control of temperature, pH, dissolved oxygen and CO_2 concentration. Homogeneity of the culture environment is controlled with agitation, while other relevant operating parameters such as osmolality and nutrient concentration are controlled via media formulation, feed addition or, in the case of continuous systems, perfusion rate.

In the following, we first describe the smallest scale-down models – for example, in the nanolitre scale – with very high throughput. Next, we move to larger models but with smaller throughput and proceed to the most common scale-down model – that is, the bench-scale bioreactor – say, in the one-litre scale. Limitations and opportunities to use them as modelling or screening tools for perfusion processes are discussed with specific attention to the problem of reproducing continuous media exchange and cell retention.

3.2 NANOFLUIDIC CHIPS

In Chapter 1, we introduced the challenges encountered in clone screening, when establishing a new manufacturing cell line. Berkley Lights has developed a fully integrated platform (Beacon) that uses Opto Electronic Positioning technology to perform and manipulate cells on nanofluidic culturing chips (Breinlinger et al. 2018). Common cell culture tasks can be programmed through a dedicated software, which also enables parallel maintenance and analysis of thousands of cell lines. This technology allows multiplexed deposition of single cells into an array of individual chambers with nanolitre volumes. An entire chip can be imaged in 8 minutes – including artificial intelligence-based cell counting, monitoring of clonality and growth of all clones in combination with scoring of antibody secretion by fluorescent assays. With this, the relative protein productivity can be estimated, and only clones with acceptable productivities can be selected (Goudar et al. 2017). This technology has not yet been implemented in industry for screening clones in perfusion mode, but, in principle, continuous media exchange could be applied using the constant medium flow along the channel. The interface between this laminar flow and the well would enable some exchange to replenish nutrients and evacuate toxic waste metabolites. It has been shown that this technique not only reduces the workload for the establishment of clonal cell lines but also allows effective clone selection for all kind of production processes. In particular, this system allows the selection of cells with rare phenotypes or desirable traits, such as high productivity for a fed-batch process or steady growth and productivity for long-term perfusion runs (Le et al. 2018).

This technology is very specific to cell line screening, but it is discussed here because it is very innovative, and it is the only known device that could allow the cultivation of single cells with a continuous medium exchange at this scale.

3.3 INCUBATED DEVICES AS SCALE-DOWN MODELS

In this section, we consider culture devices such as Erlenmeyers, shake tubes or deep-well plates that are incubated in a controlled environment able to maintain viable culture conditions. Typically, temperature, humidity and CO_2 are controlled in the incubation environment itself and agitation is used to favour mixing and oxygen transfer to the culture.

These culture devices are closed to maintain sterility, and a gas-permeable membrane allows for oxygen and CO_2 transfer. The oxygen transport from air into the liquid suspension, where it is consumed by the cells, is controlled by the agitation in the liquid phase and the surface area of the gas–liquid interphase. Since these are both modest, oxygen supply in these devices may become a problem. Pure oxygen is sometimes used in the sparger line of benchtop bioreactors, but it is not an option for incubation devices because of safety reasons. In these devices, pH is neither monitored nor controlled, but a buffer system based on the equilibrium of sodium bicarbonate and CO_2 in air is used. The following equations describe the buffer system of sodium bicarbonate decomposing into weak acid carbonate and hydroxide while remaining in equilibrium with CO_2:

$$2NaHCO_3(s) + H_2O(l) \rightleftarrows 2Na^+(aq) + HCO_3^-(aq) + OH^-(aq) \quad (3.1)$$

$$HCO_3^-(aqu) + OH^-(aq) \rightleftarrows 2OH^-(aq) + CO_2(g). \quad (3.2)$$

It is also worth noting that the removal of CO_2 from the culture suspension is not very efficient in these devices, since the stripping effect – characteristic of the air bubbles rising in a benchtop bioreactor – is missing. On the other hand, agitation is generally sufficient to ensure efficient heat transfer and satisfactory homogeneity of the cell suspension. The sample strategy has to be designed depending on the working volume, which may differ significantly among the different devices, as described in detail in the following section.

3.3.1 Shake Tubes and Deepwell Plates

Shake tubes were developed to serve as scale-down models for mammalian cell culture (De Jesus et al. 2004). They are typically 50 mL tubes designed to be used in incubated devices with agitation (Figure 3.1). The specific gas–liquid interphase surface area can be optimised by changing the culture volume in the tube and also the agitation conditions. The interphase surface area in the tube increases when the agitation is increased (Zhu et al. 2017).

Operations like sampling, seeding, feeding, etc., are done manually by an operator under a laminar flow hood to guarantee sterility. The culture volume of these models is sufficient for thorough characterisation of the cell culture in terms of cell density, metabolite concentration, osmolality, spent media analysis, product quantification

Figure 3.1 Shake tubes of different size. The white spots on the yellow caps correspond to gas permeable membranes that keep the system sterile while maintaining in equilibrium the gas phases in the incubation chamber and inside the tube.

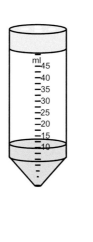
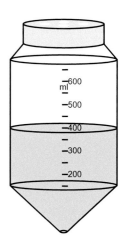

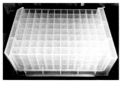

Deepwell plate set-up for incubation   Robotic platform for liquid handling

Figure 3.2 On the left-hand side, a 96-deepwell plate without the lid (top) and an incubator with the plate covered by the metallic lid (bottom). On the right-hand side, a robotic platform for liquid handling. This platform must be placed under a laminar flow hood to work under sterile conditions. Biomek® i7 Automated Workstation © 2017 Beckman Coulter, Inc. All rights reserved. Used with permission

and product quality evaluation. On the other hand, pH and DO are not monitored or controlled during the culture.

Besides the shake-tube technology, deepwell plates, which are very similar to traditional well plates but are deeper and therefore offer a larger culture volume, represent a promising alternative for the application in incubation devices (Figure 3.2). They exist in different formats (6–96 wells) and can be used to grow cells in suspension, similarly to shake tubes. During cultivation, the plates are covered with a lid that is permeable to gases but serves as a sterility barrier. By placing the deepwell plate into an incubator, agitation, pH and oxygen transfer are controlled just like in shake tubes. The working volume and the agitation speed must be carefully selected to guarantee sufficient oxygen transfer through the gas–liquid interface and good mixing conditions (Duetz 2007, Running & Bansal 2016).

With such a large number of experiments running in parallel, manual handling is not feasible, and robotic platforms have to be used for liquid handling. These devices can be programmed to sample, seed and feed each plate. For example, in a 96-deepwell plate, different media can be blended independently in each well, thus creating 96 different media composition running in parallel in the same experiment (Jordan et al. 2012, Sokolov et al. 2017a). Different robotic arms have either 96 tips used, for example, to sample the entire plates at once or independent tips that can be programmed to add/remove liquids to only specific wells and with specific amounts. These devices provide, of course, a very high throughput, and are therefore best suited for media blending and screening applications. However, the reduced volume, especially in the 96-well format, may pose limitations in terms of analytics. Quality attribute characterisation often requires a substantial amount of protein that cannot always be obtained in such small volumes. If the product concentration is high enough, one can sacrifice one well by withdrawing the entire volume and then stopping the well. This is mostly done at the end of a batch or fed-batch experiment.

Many applications of deepwell plates can be found in literature. Most of them include media blending studies, such as the one of Jordan et al. (2012) on amino acid composition, where up to 192 different conditions were generated and analysed. Rouiller et al. (2013) generated 376 different conditions starting with 16 medium formulations and shuffling 43 out of the 47 media ingredients. With respect to cell line screening, it was shown that the deepwell-plate platform can be used to predict productivity and even quality attributes at larger scale, by starting with hundreds of clones and performing selection in parallel with scale-up from shake tubes, to micro-scale bioreactors and eventually to bench-scale bioreactors (Bielser et al. 2019b, Rouiller et al. 2016). Investigations on product quality were also reported (Brühlmann et al. 2017a, 2017b; Sokolov et al. 2017a). A schematic representation of the experimental workflow in these systems is shown in Figure 3.3.

The adopted analytical devices must also be adapted to this high-throughput format. For cell counting, for example, various commercial devices exist based

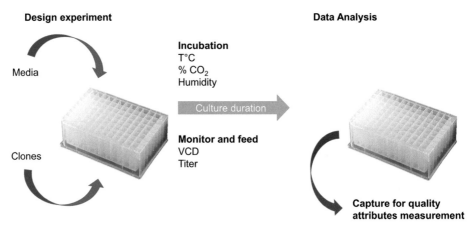

Figure 3.3 After proper experiment design, the defined amounts of media and the selected clones are added to each well of the plate. During cell culture, the plate is incubated and the robotic device is used to sample and feed the culture. High-throughput cell counting and protein titration devices are used to collect the data. At the end of the culture, the proteins can be captured for further quality attribute measurements.

on flow cytometry technologies. For the quantification of the secreted protein, high-throughput solutions are available based on fluorescent labelling or, in particular for the case of monoclonal antibodies, on epitope binding kinetics detection.

It can be noted that the format of these systems, with the handling of very small quantities, inevitably generates non-negligible variability in the final data. Therefore, we recommend using these technologies mainly for screening or performance-ranking purposes.

3.3.2 Strategies for Perfusion Cell Culture in Incubated Systems

Shake tubes and deepwell plates are well-established systems in process development for batch or fed-batch processes (Barrett et al. 2010, De Jesus et al. 2004, Gomez et al. 2017, Joao De Jesus & Wurm 2013, Rouiller et al. 2016). As mentioned in Section 1.3.3, semi-continuous operation with periodic and controlled media exchange can be used to mimic perfusion conditions in such devices. The main steps are to separate the cells from the culture suspension – for example, through sedimentation or centrifugation – and then to remove the supernatant or harvest media. This mimics the harvest stream in a real perfusion system since it contains spent medium, toxic waste metabolites and the secreted product. The removed harvest is then replaced by fresh medium that was previously conditioned, thus mimicking the perfusion stream.

Villiger-Oberbek et al. (2015) used this procedure to reach cell densities of 50×10^6 cells/mL and sustain a long-term culture for more than 58 days at about 30×10^6 cells/mL. Henry et al. (2008) predicted some parameters of the culture perfusion – such as cell growth, nutrient consumption, metabolite production and titre – using spinner flasks operated in the semi-continuous mode. Bielser et al. (2019b) used 96-DWP and ST to screen clonal cell lines, identify the best producer and predict important perfusion parameters such as CSPR and volumetric productivity. A comparison of cell-specific growth, glucose consumption, ammonia and monoclonal antibody production rates in semi-continuous shake tubes and benchtop bioreactors is discussed in detail in Section 4.4.1 (Wolf et al. 2019a). These scale-down models of perfusion cultures offer very interesting alternatives to bioreactor systems due to their much larger throughput – which makes them attractive, particularly during screening and earlier stages of process development. Of course, there are also important limitations with respect to the semi-continuous nature of the operation. For example, very high cell densities might be difficult to maintain because the required media exchange frequency might be too high. The most reasonable exchange frequency, which – keeping constant the exchange volume – is proportional to the perfusion rate, is in fact limited to once a day. Twice would require a 12 h interval, posing some organisational issues. Oxygen transfer limitations may occur in both ST and DWP systems at sufficiently large cell densities. Also, pH is not well regulated in these systems, and this can alter the product quality or even the overall culture performance (Bielser et al. 2019b, Wolf et al. 2019a). Thus summarising, although these scale-down models for perfusion are far from perfect, they can still be used not only for screening purposes but also for estimating at least a first guess of the fundamental operating parameters, such as an initial guess of the minimum CSPR, for larger scale perfusion bioreactors. Examples of applications of these systems in clone screening and process development are discussed in Chapter 4.

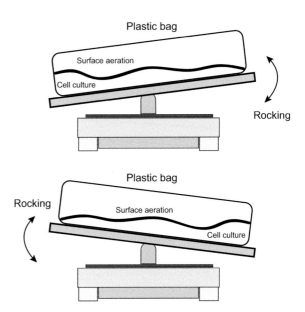

Figure 3.4 Principle of a wave bioreactor: the rocking movement creates waves and turbulences that favour oxygen transfer and mixing of the cell suspension. The bag is inflated with the desired gas mixture (often air and CO_2).

3.4 WAVEBAGS

Wavebag bioreactors are very traditional in cell culture and provide a category by themselves. Their name is self-explanatory: they are bags filled with culture media, lying on a flat surface that is continuously oscillating between two angles, as shown in Figure 3.4. Their operation is more similar to an incubated device than to a controlled bioreactor. The surface underneath the bags is heated and kept at a temperature value regulated through a thermometer located close to the bag. The gas mixture (air and 5 per cent CO_2 as a pH buffer) is introduced to the bag through a permeable filter. Because of the large surface-to-volume ratio, this usually provides sufficient exchange of the gaseous species.

Although wavebags have been used for perfusion cell cultures in the past, they are mostly used for seed train rather than for process-development-related activities. For example, the use of highly concentrated 5 mL cell bank vials, with densities of 90 to 100×10^6 cells/mL (Tao et al. 2011) in the seed train, instead of more traditional low-density banks, can save up to 9 days of expansion. This is extremely important from an operational point of view, since it reduces the time interval between thawing and inoculation of the production bioreactor. Wavebag bioreactors are ideal for generating high-density cell banks because the volume that they can handle is enough for a large number of vials, and this step does not require a large experimental throughput.

Some applications of wavebags for perfusion culture operation have been reported in the literature using tangential flow filtration (either TFF or ATF). Cell densities up to 130×10^6 cells/mL have been reported to stay viable for up to 20 days, with an impressive peak at 214×10^6 cells/mL (Clincke et al. 2013a). It is interesting to note that these results are made possible through the high gas–liquid mass transfer rates characteristic of these systems, which comes from the suspension surface-to-volume ratio that is very large in comparison to other cultivation devices. Foam formation is also not common in these devices. However, wave bioreactors are currently not considered for industrial production since the rocking movement that provides agitation of these bags limits their scale-up as the energy required must be sufficient to displace the entire culture volume, which is not feasible when considering thousands litres bags.

3.5 BIOREACTORS

A typical bioreactor system is constituted by a closed fermentation vessel and includes various devices for proper monitoring and control of the culture environment, as introduced in Chapter 2. Sterility can be preserved because the reactor is closed and all operations are performed so as to prevent contamination. To be sterile in the first place, the reactors are either autoclaved or gamma irradiated, depending on whether they are glass or single-use plastic devices. Different types of sensors monitor temperature, pH and pO_2 while control loops are in place to keep these parameters within specified intervals. In most cases, pCO_2 and other more sophisticated parameters, such as the biomass or specific nutrient levels, are also monitored and controlled. In perfusion systems, the flowrates of the media inlet, the harvest and the bleed stream are also subject to different control strategies, as already discussed in Chapter 2. In the following, we describe how the different parameters are controlled in the context of a scale-down model for the manufacturing scale.

3.5.1 Automated Microscale Bioreactors

So-called microscale bioreactors are among the smallest available scale of bioreactors on the market. They exist at two different scales, 15 and 250 mL working volume, and are operated in blocks of 24 or 48 reactors for the 15 mL scale and 6–24 for the 250 mL scale. A robotic platform can be programmed by the operator for different tasks, such as sampling and feeding, but also to monitor and control the set points of the different culture parameters. The platform can perform liquid handling using a robotic arm and sterile tips, gas lines are directly connected to the bioreactors, agitation is provided via an impeller, and the stations on which the bioreactors are located can be temperature-controlled. The mL working volume is comparably high compared to the 96-DWP or the nanofluidic chip technology, and therefore, both shake-tube and ambr® systems are better suited for the early stages of process development than for screening applications.

Examples of these devices are shown in Figures 3.5 and 3.6, where the vessels are in single-use format and include pH and DO monitoring probes. The advantage of micro-scale bioreactors is obviously that the experimental throughput is significantly increased compared to traditional bench-scale bioreactors. This is achieved also through a high level of automation in sampling and feeding, which does not require any human intervention.

The ambr® 250 is a very recent device, but it has already proven to be an efficient scale-down model. Xu et al. (2017a) report that, using the rate of gas–liquid mass transfer as a scale-down parameter, they could generate comparable cell growth and protein production profiles in the ambr® 250, in 5 and 1,000 L bioreactors.

In both of the equipment described in the previous paragraph, monitoring of cell density, metabolite concentrations and quality attributes is done through offline or at-line sampling. For this, the 15 mL format may not be suitable, due to volume limitations and, particularly in the case of monitoring quality attributes over time, the

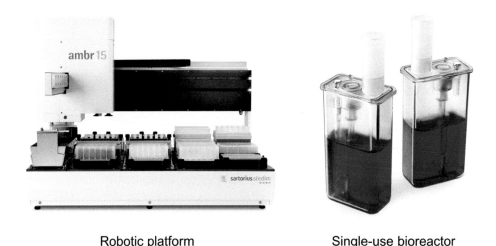

Robotic platform   Single-use bioreactor

Figure 3.5 The ambr® 15 high-throughput station for 24 bioreactors (left) and schematic of the 15 mL single-use bioreactor designed for this platform, ambr® images provided courtesy of Sartorius Stedim Biotech

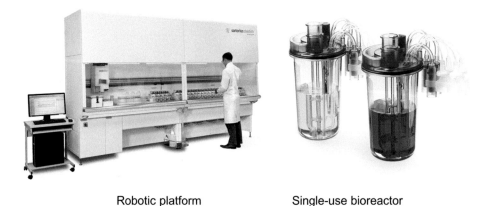

Robotic platform   Single-use bioreactor

Figure 3.6 The ambr® 250 high-throughput station for 24 bioreactors (left) and the 250 mL single-use bioreactor used in the platform, ambr® images provided courtesy of Sartorius Stedim Biotech

sampled volumes may become significant with respect to the feed volumes. Therefore, when planning a series of experiments, like in the case of the design of an experimental program, the sample volumes required during cell culture must be verified to ensure that the maximum and minimum values are within the ranges provided by the respective equipment.

Although these systems were originally designed to mimic batch and fed-batch operations, the extension to perfusion cultures, by introducing continuous media exchange and cell retention is possible although not simple. In these platforms, the bioreactors are fixed in agitation blocks that hold twelve bioreactors each. This makes handling the bioreactors for semi-continuous operation cumbersome. However, Janoschek et al. (2018) used centrifugation, similarly to what was described earlier for shake tubes and deepwell devices, to maintain steady-state conditions at a density of 30×10^6 cells/mL for a triplicate and for about one week. An alternative, in order to replace spent media with fresh media without removing cells, would be

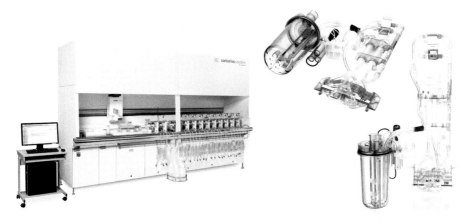

Robotic platform   Single-use bioreactor and hollow-fibre module

Figure 3.7 Robotic platform and single-use bioreactors for perfusion cell culture on ambr® 250. The bioreactors are connected to hollow fibres that can be operated either as TFF or ATF. Ambr® images provided courtesy of Sartorius Stedim Biotech

to let these cells settle in the bioreactor. For this, agitation and aeration should be stopped for some time until the cells settle (Davis et al. 2015, Goletz et al. 2016, Kreye et al. 2015). This can certainly lead to problems with respect to oxygen requirements, particularly in the case of higher cell densities. Nevertheless, the automated mini-bioreactor (ambr®) system at the 15 mL scale can be very convenient due to the high degree of automation that allows programming the sedimentation and medium exchange steps with a dedicated software. The system allows the operation at medium exchange rates of up to two reactor volumes per day and sedimentation yields cell separation efficiencies above 95 per cent.

Recognising the need for a real perfusion device, a novel type of bioreactors was developed for the ambr® 250. These bioreactors are equipped with hollow-fibre modules, as shown on Figure 3.7, and that can be used either in the TFF or in the ATF flow mode (Xu et al. 2017a, Zoro & Tait 2017).

3.5.2 Bench-Scale Bioreactors

Bench-scale bioreactors are glass or single-use plastic vessels in the volume range of one to ten litres. This scale is often used as a reference point for the scale-up to clinical and commercial scales, because the processes occurring at this scale are typically sufficiently similar to the corresponding ones occurring at the larger scale to allow for a reliable scale-up based on well-known engineering procedures. In a typical laboratory set-up, a central control unit receives data and sends instructions to a number of peripheral process units. Each of these controls a single vessel (Figure 3.8). Online monitoring and control typically include a temperature sensor, a pH probe, an optical dissolved oxygen probe and sometimes a dissolved carbon dioxide probe. The online data generated by these sensors are processed by the central control unit that, in turn, operates on the different process parameters.

Temperature is controlled through a heating jacket surrounding the vessel, operated through a water circuit heated externally, or an electric resistance device. In both

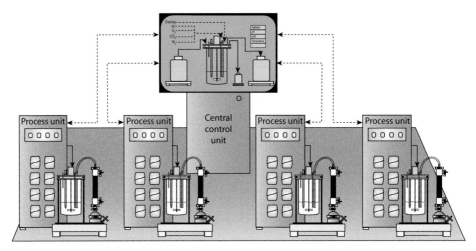

Figure 3.8 Schematic of the series of four bioreactors with a central control unit that collects online measurements and drives the different peripheral units located behind each bioreactor. These distribute the different services, such as heating/cooling water, gases, stirring and feeds or media, to the bioreactors.

cases, an efficient heat exchange requires that the liquid inside the vessel is sufficiently well agitated. This is typically the case in mammalian cell cultures since sufficient mixing is needed anyway to keep the cells in suspension and ensure homogeneity. Mixing is indeed a crucial point also considering its role on the rate of gas–liquid mass transfer. On the other hand, the shear generated by mixing can be dangerous for the cell viability. This aspect is more critical for mammalian cells than for other cell types, like bacteria or other microorganisms, due to their more delicate external membrane. However, at the scale of a few litres, a good compromise among all these factors can easily be found, but the situation becomes more critical at larger scales. For this reason, the issue of providing sufficient mixing through a proper design of the stirrer geometry and speed, without endangering the cell viability, is treated in more detail in Chapter 5.

The proper control of the bioreactor pH is essential since it impacts the cell culture viability, the cell-specific productivity and even the product quality such as its glycosylation pattern (Ivarsson et al. 2015; Monteil et al. 2016; Villiger et al. 2016a, 2016b; Yoon et al. 2005). For this, we need to have the possibility of taking two different actions: one to increase pH and the other to decrease pH. In particular, a base solution (sodium bicarbonate or sodium hydroxide) is commonly used in the first case, while the second action is performed by injecting CO_2 in the reactor, which acts on the pH through the following chemical reaction:

$$CO_2(g) + H_2O(l) \rightleftarrows H_2CO_3(aq) \rightleftarrows HCO_3^-(aq) + H^+(aq) \tag{3.3}$$

The pH is also affected by the metabolism of the cells, as they might produce or consume acidic/basic species over the course of the culture (Li et al. 2012, Mulukutla et al. 2012, Zagari et al. 2013).

On the other hand, since pH is such a sensitive indicator of the cell culture state, it can also be used to control the reactor operation. For example, Gagnon et al. (2011) used the pH value to control the glucose concentration in the reactor by acting on its feed rate. Xu & Chen (2016) reported instead the case of a cell culture that was sustained at high density and over several days without close pH control. This is probably a special feature of the specific cell line and media considered, which does not alter the conclusion that pH needs to be closely monitored and controlled in bioreactors.

Dissolved oxygen concentration in the cell culture is crucial for at least two reasons. Firstly, oxygen is consumed by the cells in the tricarboxylic acid cycle (TCA) cycle (respiration), which then cannot survive without oxygen. Secondly, high concentrations of dissolved oxygen may favour the generation of reactive oxygen species (ROS) and harm the viability of the cell culture. Therefore, the saturation of dissolved oxygen in the liquid phase of a mammalian cell culture has to be controlled at a prescribed percentage (for example 50 per cent of saturation with air). This guarantees that enough oxygen can be supplied to the cells for respiration but minimises the formation of reactive oxidating species. Dissolved oxygen can be controlled by a stream of air or pure oxygen to the culture, which keeps its level within the desired bounds.

The rate of oxygen transfer to the liquid phase depends on different factors and is quantified by the volumetric mass transfer coefficient known as $k_L a$, given by the product between the gas–liquid mass transfer coefficient, k_L, and the interphase surface area per unit reactor volume, a, multiplied by the driving force, related to the concentration difference between the liquid and the gas phase (Tribe et al. 1995). This accounts for the effect of many variables on the gas–liquid mass transfer rate. For example, going from air (≈ 21 per cent O_2) to pure oxygen increases the driving force of mass transfer, while the shape of the sparging pipe affects the interphase surface area. The latter is a consequence of the effect of the sparger type, ranging from open pipes to drilled holes or even micropores, where the size of the sparging element can vary from a few µm to a few mm, on the bubble size. This has obvious consequences on the interphase specific area which, for example, increases by 10 times if, for the same volumetric gas flow, the bubble diameter decreases from 1 mm to 0.1 mm. It is worth noting that, before inoculation or in general at very low cell densities, the gas consumption rate is very low and, therefore, the oxygen concentration in the liquid phase approaches equilibrium conditions with the gas stream. If this is too large, it can be decreased by properly diluting the gaseous stream with a second one made of pure nitrogen.

Aeration using spargers is efficient with respect to gas exchange but also has some disadvantages. When the bubbles reach the surface of the liquid, they burst, thus locally delivering some energy in the form of shear that can damage the cells. Usually, the smaller the bubble is, the larger the energy liberated and the larger the damage to the cells (Chisti 2000, Meier et al. 1999, Xu et al. 2017c). But, as discussed earlier, small bubbles are more efficient for oxygen transfer and therefore generally preferred in the case of high-cell-density bioreactors. As seen in Section 2.5, some media components such as pluronic acid can act as shear protectant for the cells and therefore strongly reduce this issue. Foam is a second side effect of large aeration rates and small bubbles. This is particularly disturbing at the small scale, where it accumulates, quickly reaches the upper region of the vessel and enters the condenser. This is, in fact, used to retain the humidity of the gaseous stream leaving the liquid suspension inside the reactor. When too much foam accumulates in the reactor, it enters the condenser and eventually reaches the filter located at its outlet. This can result in filter clogging and, as a consequence, the pressure inside the vessel increases, causing the interruption of the operation. Different strategies can be adopted to counter-act excessive foam formation. These include commercial polymers that modify the surface tension of the bubbles (Eleftherios 1991, Routledge 2012) and are typically added from the top of the bioreactor to prevent foam formation. The concentration of these compounds must be closely controlled because they can have some toxic effects. An alternative is the introduction of classical physical foam-breaking devices (Deshpande & Barigou 1999, Ishida et al. 1990, Takesono et al. 2006). At the large scale, foaming is less of an issue: sprinkler devices can be used to spray the culture surface and break the foam. The added water volume is negligible compared to the reactor volume.

3.6 SENSORS

3.6.1 In Process Control

The operations performed to periodically sample and characterise offline various parameters in the bioreactor are usually referred to as IPC. For example, viable cell density, pH, and metabolites like glucose and lactate are quantified based on a suitable sample taken on a daily basis. The obtained pH value is then used to correct possible deviations from the set-point value, while the information about cell density and viability serves to observe the progression of the culture. Metabolite concentra-tions like glucose, glutamine, glutamate, lactate, and ammonia are also commonly used to monitor the process. In particular, from the glucose concentration the feed addition needed to prevent its full depletion is calculated. Obviously, the methods used for offline monitoring of cell cultures are not affected by the operation mode, whether batch or continuous. We therefore do not provide here a description of the many possibilities in this direction and refer the interested reader to the specialised literature (Hu 2012, Zhou & Kantardjieff 2014).

Compared to online monitoring and control, IPC – with its daily sampling, sam-ple preparation and characterisation – is obviously far more time consuming and less efficient, although in general can, of course, access more characteristics of the culture and the involved biomolecules. For example, considering the quantification of the metabolites, the sample needs to be first centrifuged and filtered. Other quan-tifications of the spent media (amino acid or vitamin concentrations for example) and critical product quality attributes (N-linked glycosylation, fragments, aggregates and charge variants) are usually performed in a single sequence only at the end of the culture run. In these cases, in fact, sample preparation takes significantly more time and cannot be performed on a daily basis. A non-exhaustive list of parameters that are measured during a typical bioreactor run and the corresponding needed sample preparation are reported Table 3.1.

More advanced online monitoring using spectroscopy techniques do exist and are continuously being developed. These can potentially replace some of the previously mentioned daily or end-of-run methods, leading to faster and more reliable process monitoring. Some of these methods are discussed in the following section.

3.6.2 Spectroscopic Sensors

Spectroscopy, based on the interaction between matter and electromagnetic radi-ation, provides the foundation for a number of analytical techniques using a

Table 3.1 Parameters typically monitored during a cell culture run. Their frequency can be very different and range from online monitoring to daily sampling, to end-of-run characterisation.

Target	Parameter	Frequency	Sample preparation
Bioreactor monitoring	T, pO_2, pH, pCO_2	Online	No sample
Cell quantification and health	VCD, viability, cell diameter	Daily	Possibly dilution
Reactor environment	pH, pCO_2	Daily	Sample
Metabolites	Glucose, lactate, glutamine, glutamate, Na^+, K^+	Daily	Filtration (0.22 μm)
Spent media analysis	Amino acids, vitamins, trace elements	End of run	Filtration (0.22 μm) + dilution
Product quality	Charge variants, N-linked glycosylation	End of run	Filtration (0.22 μm) + dilution + glycan isolation
Contaminants	Fragments, aggregates	End of run	Filtration (0.22 μm) + dilution

wide range of frequencies. These include ultraviolet (UV), near-infrared (NIR), mid-infrared (MIR), Raman and fluorescent spectroscopy that can be used to obtain spectral characteristics of different materials, from which information about atomic, molecular or even crystal structures of the material can be derived. Some of these techniques and the corresponding wavelengths are presented in Figure 3.9.

Different spectroscopy techniques were used to monitor online mammalian cell cultures either to measure a single quantity or, coupled with chemometric modelling techniques, multiple quantities simultaneously. Focus is given here on a few of these techniques that seem to be most promising, by discussing their potential and limitations. Most of these techniques are not yet mature enough for routine application, but due to the high potential to develop increasingly efficient process analytical technologies (PAT), particularly in the context of continuous biomanufacturing, it is not difficult to predict that they will strongly grow in the near future (Swann et al. 2017). Currently, in most cases, only temperature, dissolved O_2/CO_2 and pH are controlled online. However, using these monitoring techniques, this situation could change. For example, the feed stream could be prepared online and on-demand, based on some advanced monitoring techniques.

It is worth mentioning that the amount of information contained in such online data is very large. More complex real-time data treatments, based on multivariate data analysis (MVDA), possibly coupled with suitable mechanistic models, are needed to extract information to be directly implemented in reactor monitoring (Glassey et al. 2011). In the following, we focus on the most promising techniques with respect to continuous cultures.

Dielectric Spectroscopy. Dielectric spectroscopy, in contrast with the techniques illustrated in Figure 3.9, is based on the interaction of an external field with the

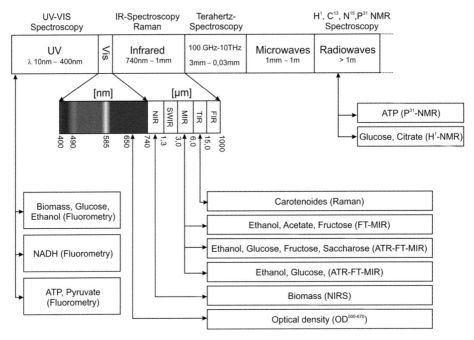

Figure 3.9 Electromagnetic wavelength spectrum and ranges of the corresponding optical and spectroscopic analytical techniques. © Reprinted by permission from *Springer Nature: Springer, Applied Microbiology and Biotechnology*, 'In situ sensor techniques in modern bioprocess monitoring', Beutel S. and Henkel S., 2011

electric dipole moment of the sample; it was used to quantify the biomass even decades ago (Ducommun et al. 2001a, Joeris et al. 2002, Konstantinov et al. 1994). This technique is based on the dielectric interaction of the cells with the surrounding media as a function of frequency (Davey et al. 1993). Since the viable cells are polarised, their membrane retains some of the ions moving towards the electrodes, leading to a polarisation potential that is measured as a capacitance (pF) value. To be independent of the geometry of the probe, the capacitance is normalised into a permittivity value (pF/cm).

Only viable cells are detected using this technique, because the membrane of cells that are undergoing apoptosis or necrosis is porous and cannot be polarised. Therefore, permittivity is affected by the viable biomass, which is related to the VCD through the cell diameter d (assumed to be spherical) as follows:

$$\text{Biomass} = \frac{1}{6}\pi d^3 \times X_V, \qquad (3.4)$$

but also by many other parameters in a cell culture that influence its dielectric characteristics, including some physiological changes that the cells may undergo (Opel et al. 2010, Zalai et al. 2015). Nevertheless, these sensors, proved to be very efficient to monitor viable cell densities online. This is done through the development of suitable correlations between permittivity and viable cell density, based on VCD measurements taken offline by a cell counter. In cases where the cell metabolism and, in particular, the diameter are rather constant throughout the culture, these correlations are linear, as illustrated in Figure 3.10.

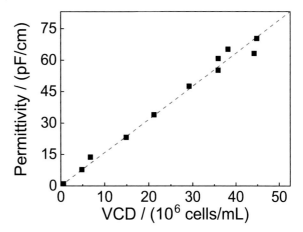

Figure 3.10 Linear correlation of the permittivity signal with the offline measured viable cell density.

Dielectric spectroscopy has been used to control fed-batch bioreactors, as reported by Ansorge et al. (2010), Zalai et al. (2015) and Zhang et al. (2015a). The feed flow rates were adapted to the cell growth rate not only to reduce inter-run variability but also to improve the performance of the cultures. For perfusion cell cultures, this sensor is often used to control the bleed rate, as described by Dowd et al. (2003), Karst et al. (2016) and Warikoo et al. (2012). This real-time reaction to changes in the cell culture behaviour allows reaching a stable biomass concentrations faster.

However, it is worth pointing out that keeping constant permittivity means keeping constant the overall biomass in the reactor. This is equivalent to controlling the viable cell density only if the cell size remains constant, and, therefore, the correlation between the two, such as the linear correlation shown in Figure 3.10, stays unchanged throughout the entire culture.

The following example provides a very clear illustration of this concept (Bielser et al. 2019b). Two clones, producing the same monoclonal antibody, were cultivated in a 26-day long-term perfusion run by maintaining a constant permittivity signal in the bioreactor. In the two runs, the cells were inoculated at a similar VCD of 0.5×10^6 cells/mL (Figure 3.11A). The perfusion was started after 3 days at a targeted value of 1.0 RV/day, and the bleed was activated as soon as the permittivity set point of 70 pF/cm was reached (Figure 3.11B), which happened for both clones at day 11. For Clone 1, the VCD at this point was around 63×10^6 cells/mL, while for Clone 2, it was around 40×10^6 (Figure 3.11A). On the other hand, the cell diameter was equal to 14.7 µm for the first one and 16.5 µm for the second (Figure 3.11C). Note that both these values have been measured offline, and from these, the biomass percentage in the reactor shown in Figure 3.11B has been computed using Equation (3.4). Thus, we see that having reached the same permittivity value, the two cultures exhibit the same biomass, although different VCD (and cell diameter) values. After day 11, as expected, due to the control action, the permittivity remained constant 70 pF/cm for both clones. However, the behaviour of the VCD was significantly different. In particular, for Clone 1, the VCD decreased slowly from 63 to 42×10^6 cells/mL between day 11 and 17 and then remained stable for the rest of the culture. For Clone 2, instead, the VCD remained constant and stable at around 40×10^6 cells/mL until the end of the culture. Similarly to the VCD, the behaviour of the cell size was also different for the two clones. For Clone 1, the cell diameter increased from 14.7 µm

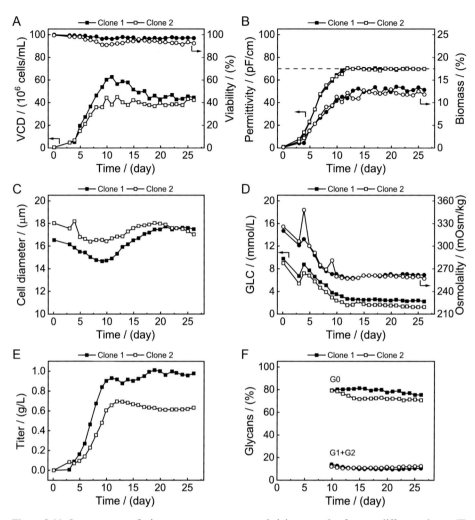

Figure 3.11 Long-term perfusion run at constant permittivity set point for two different clones (■ Clone 1, □ Clone 2) : (A) VCD and viability, (B) Permittivity, permittivity set point (red dashed line) and calculated biomass using Equation (3.4), (C) Cell diameter, (D) Glucose concentration and Osmolality, (E) titre and (F) Glycans G0 and (G1+G2) as a function of time. © Reprinted from *Biotechnology Progress*, Jean-Marc Bielser, Jakub Domaradzki, Jonathan Souquet, Herv Broly and Massimo Morbidelli, 'Semi-continuous scale-down models for clone and operating parameter screening in perfusion bioreactors', 2019, with permission from John Wiley and Sons

at day 11 until it reached a plateau value of about 17.5 μm (Figure 3.11C), while for Clone 2, a similar behaviour was observed, with a minor oscillation of the cell diameter between 16.5 and 17 μm. In both cases, as shown in Figure 3.11B, the computed value of the reactor biomass remained substantially constant, as expected from the constant values of the controlled permittivity.

Interestingly, the values of glucose concentration, osmolality, mAb concentration and product quality (i.e., glycan patterns) in Figure 3.11D–F remain constant during the entire culture after day 11. This shows that for both clones, the total biomass-based control succeeded in keeping stable the bioreactor steady state, independently of the behaviour of the VCD, which was, in fact, clone specific.

The preceding experiments show, in particular, that we have to distinguish between at least two types of cell behaviours. In the first one, as for Clone 2, the cell diameter is independent of the state of the cell or shows only slight changes after reaching the targeted permittivity/biomass set point and entering the production phase. In this case, no differentiation between viable cell density and biomass is necessary, and the operation at steady state by total biomass control is possible. In the second type of behaviour, as for Clone 1, the diameter is still changing, typically increasing for some days after the total biomas has reached the stationary phase. As a result, constant biomass does not imply constant VCD, and then steady-state operation becomes questionable. However, at least for the case investigated here, operation at constant biomass still enables constant metabolite and mAb levels, as well as product quality (Figure 3.11D–F), independently of the changes in the VCD value. Nevertheless, the fact that the cell diameter is changing indicates that steady-state conditions have not actually been achieved since some evolution in the cell metabolism is still apparently taking place.

Raman Spectroscopy. Raman spectroscopy is based on the inelastic scattering of monochromatic light directed at the analyte, resulting in a wavelength shift. Although only a small portion of the photons are inelastically scattered – i.e., about one out of 10^9 or 10^{10} – these carry a lot of information and, therefore, provide a true spectral fingerprint of the analyte that can be used for identification and quantification (Buckley & Ryder 2017). Typically, a Raman spectrograph includes three components: a laser, which provides a high-intensity monochromatic source of light, several sampling optics and a detector enabling the acquisition of Raman spectra (Atkins et al. 2010, Esmonde-White et al. 2017).

Raman spectroscopy is a non-destructive and *in situ* analytical method that does not require sampling. It is a robust technique that, in principle, may enable monitoring multiple quantities at the same time. Moreover, it provides important online process information. This last characteristic is probably the most attractive one, also, in the case of mammalian cell cultures.

The main difficulty in using this sensor is related to the extraction of the desired information from the measured spectra. In a clean sample, with few and simple molecular structures, distinct peaks can be observed in the spectra, and each can be directly related to the sample composition. However, in complex samples, such as in the case of a cell culture, which contains a large number of Raman active chemical and biological species, the measured spectra largely consist of very broad envelopes resulting from the overlapping of the peaks corresponding to the many involved species. It is therefore necessary to reduce the high content of spectral noise and emphasise the information content. For this, various pretreatment and modelling tools based on different statistical approaches have been developed (Gautam et al. 2015). Probably the most common one is the partial least square (PLS) regression, which is described in more detail in Chapter 6. A crucial point is the calibration step, where a suitable model correlating the measured spectra with the sample composition is developed and calibrated using a proper set of offline measurements. This model is subsequently used to predict from measured spectra the composition of new samples. The selection of suitable data sets with sufficient variability is fundamental to obtain a reliable predictive model due to the intrinsically limited extrapolation capability of statistical models (Mehdizadeh et al. 2015, Opel et al. 2010, Webster et al. 2018). For example, if a model is calibrated with a specific pair of cell line and medium,

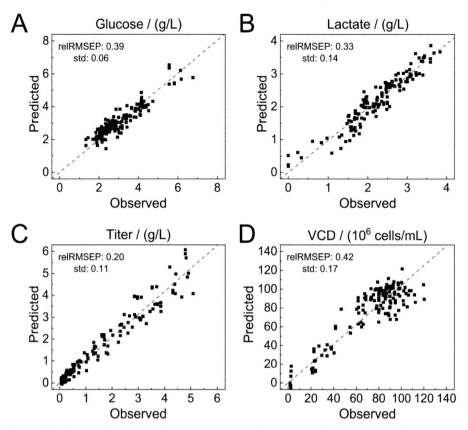

Figure 3.12 Correlation between Raman-based predictions using a PLS model and offline measurements for glucose (A), lactate (B), titre (C) and VCD (D).

it would not necessarily work for a different combination of cell line and medium. This is clearly a limitation of this technology, especially in a process-development environment that is exposed to a large variety of cell lines and media.

Nevertheless, there are several successful implementations of Raman technology reported in the literature. It has been used not only for raw material and media characterisation but also for monitoring and controlling various parameters during cell culture (Buckley & Ryder 2017, Goh et al. 2017). These include the concentration of various metabolites such as glucose, glutamine, glutamate, lactate and ammonia as well as the viable/total cell density (Abu-Absi et al. 2011, Webster et al. 2018, Whelan et al. 2012). With respect to process-control applications, Berry et al. (2016) adjusted the semi-continuous glucose feeding rate based on the online glucose measurement through a Raman sensor, with the aim of reducing undesired glycation of the produced protein. Matthews et al. (2016) used Raman spectroscopy to monitor lactate and glucose levels, and adjusted the glucose feed in order to keep lactate below a certain level while avoiding total glucose depletion. Using this method, the culture could be significantly prolonged, reaching a final titre increase by 85 per cent.

In order to illustrate the potential of this technology, the Raman-based predictions of glucose and lactate concentrations, titre and viable cell density in a representative industrial mammalian cell culture are compared in Figure 3.12 as a function of the corresponding quantities measured by classical offline analytical techniques. In this comparison, PLS-based models were used after training on a small but representative data set of six different fed-batch runs. Five runs were used for cross-validation while the sixth one was predicted, resulting in a certain relative root mean square error in prediction (relRMSEP). This was repeated through a rotational approach so that each batch run was predicted once. The obtained results are also shown in the same figure in terms of the average of the errors given by the different rotations (av-relRMSEP) and the corresponding standard deviation (STD). It can be seen that the predictions nicely group around the diagonal going through the origin, with a slight scattering, which is maximal for VCD. The obtained comparison is indeed satisfactory and proves that Raman spectroscopy has, indeed, the potential to provide online several relevant information about the behaviour of the process. It is worth mentioning that the accuracy and robustness of the model predictions can in general be improved by adding more and suitably selected data to the calibration set (Feidl 2019, Feidl et al. 2019).

A very attractive application of Raman spectroscopy in perfusion cultures could be to define online whether steady-state conditions have been achieved or not, due to its sensitivity to small variations in the bioreactor composition. A principle component analysis (PCA), discussed in Chapter 6, could be very useful in this context to quantify small spectral variations and indicate measurement patterns and process evolution. Hence, small drifts from steady state could be detected, and a proper control action could be triggered. Other potential applications include the control of the inlet and outlet reactor streams, such as the bleed by monitoring the VCD, or the control of an independent glucose feed pump dedicated to maintain very low levels of glucose concentration. The possibility of using a single spectrum, and therefore a single probe to reliably monitor several parameters at the same time, makes Raman a very promising technology that we are going to see used more and more often in process monitoring.

Near- and Mid-Infrared Spectroscopy (NIR and MIR). Infrared spectroscopy is another technique used for the quantification of multiple attributes in cell culture processes (Vojinović et al. 2006). In particular, near (NIR) and mid-infrared (MIR) refer to infrared spectroscopy operated in the wavelength ranges of 0.75–1.4 μm and 8–15 μm, respectively. NIR is easier and less costly to employ than MIR, but the latter offers a much better resolution. Because of the high absorbance of water, MIR is, nevertheless, more difficult to use.

As for Raman spectroscopy, suitable chemometric models are needed to best correlate the measured spectra with the quantities that have to be measured (Cervera et al. 2009, Scarff et al. 2006, Vojinović et al. 2006). NIR and MIR have been widely used in the literature to estimate a variety of parameters in very different types of cultures such as yeast, fungi and mammalian cells (Cervera et al. 2009). Finn et al. (2006) reported the use of NIR for the estimation of biomass, glucose, ethanol and protein content in a yeast-based process, whereas Yeung et al. (2002) used the same technology to monitor a flocculation process. Raw material screening for cell culture media was also reported (Kirdar et al. 2010).

Monitoring mammalian cell cultures using NIR or MIR is not very common, and it actually has never been demonstrated in perfusion applications. Arnold et al. (2003) and Ducommun et al. (2002) used an *in situ* probe and created suitable models to estimate glucose, lactate, glutamine and ammonia concentrations in a CHO-K1-based process. The developed calibration technique provided a reliable model for the interpretation of the raw signals but required a substantial set of offline measurements. Also, Roychoudhury et al. (2007) proposed a multiplexed calibration approach for the monitoring of glucose and lactate, again *in situ* and with mammalian CHO cells.

It is to be noted that infrared, until now, has been mostly used for bacterial fermentations, but, similarly to Raman, it also has as strong potential for online monitoring in mammalian cell cultivations.

3.7 CONCLUSION

The renewed interest in perfusion bioreactors in the context of biomanufacturing is still relatively recent, and most of the related ancillary technologies available on the market are relatively old. This explains why the scale-down models and the systems for monitoring and control of perfusion bioreactors are relatively scarce. Indeed, things are rapidly changing, and we observe a rapid growth in the availability of devices of this type, generally referred to as PAT, from various vendors. This applies also to the process of digitalisation, which is expected to enter much more into the bioprocessing sector in the coming future. This will change the way we approach the design and development of industrial perfusion bioreactors today.

In this chapter, we discussed the availability – and, actually, in most cases, the lack – of scale-down technologies for perfusion cultures. The continuous nature of perfusion and the need for cell retention make the design and realisation of equipment in the nanolitre and millilitre scales technically difficult. For this, the semi-continuous operation has been developed to mimic perfusion, at least in a discretised form. Table 1.3 in Chapter 1 summarises the devices that, mostly based on this approach can currently be used as perfusion scale-down models and their respective monitoring and control characteristics.

Online or at least at-line devices for monitoring and controlling perfusion cultures are also becoming more commonly available, in an attempt to move biomanufacturing to the next stage, dominated by automation and digitalisation. Capacitance is currently the most common variable measured online for monitoring cell biomasses. Raman spectroscopy appears to be very promising for monitoring several variable simultaneously – like titre, nutrients, metabolites and possibly some product-quality-related variables like protein aggregates or glycoforms.

Advanced online monitoring techniques are currently mostly applied at the bench scale, but great benefits could also be obtained by using them in automated micro-scale bioreactors, such as those described in this section. Coupled with the existing high level of automation, these could allow driving the process with an even smaller external interaction. Sampling for cell counts, for example, could be avoided, and the perfusion rate or the bleed rate could be controlled, based on the real-time sensor information.

4

Design and Optimisation of Mammalian Cell Perfusion Cultures

In this chapter, we discuss state-of-the-art strategies for perfusion process design, development and optimisation. We first introduce the essential steps and boundaries for the development of a perfusion process. Next, we evaluate clone and media screening for perfusion processes and compare various alternative scale-down systems. Once the expression system is selected, we proceed with the design of the reactor operating conditions to maximise volumetric productivity and yield while reducing medium consumption. The fundamental issue of product quality attributes for biologics and biosimilars and the impact of perfusion operating mode are also discussed. Finally, we introduce the main objectives in the scale-up of the developed process to clinical and commercial productions.

4.1 PRINCIPLES OF OPTIMAL PERFUSION BIOREACTOR DESIGN

The implementation of a perfusion process has to address some issues, which refer not only to production costs and product quality but also to process robustness. The most reasonable starting point is probably to select the maximum volumetric throughput while maximising the protein-producing biomass inside the bioreactor, in order to improve process economics. In Chapter 2, we have seen that a key parameter in this regard is the ratio of volumetric throughput, represented by the perfusion rate, P (RV/day), and the viable cell density, X_V (10^6 cells/mL), defined as the cell-specific perfusion rate, CSPR (pL/cell/day). The CSPR defines the amount of medium fed to a single cell per day, and therefore, reducing the CSPR represents an important factor in optimal process design (Konstantinov et al. 2006).

In the following, we discuss the robust design of perfusion cultures going through, first, the proper selection of clones and media; next, the determination of a suitably low CSPR, which in the best case is equal to the $CSPR_{min}$ value defined in Section 4.3; and, finally, the best operating conditions in terms of viable cell density and perfusion rate at the given $CSPR_{min}$.

4.1.1 The critical Bioreactor Parameters

In Chapter 2, we discussed the strategies to control the system variables – such as pH, DO, agitation, working volume, flowrates of the inlet and outlet streams, and viable cell density. The implementation of such strategies requires the definition of proper upper and lower boundaries, which are particularly relevant for viable cell density and perfusion rate or CSPR.

In the case of the perfusion rate, the upper limit is mainly determined by two factors: set-up operation sustainability and economic convenience. Indeed, the cell retention device cannot sustain too large perfusion rates and therefore imposes limitations, while, on the other hand, high perfusion rates imply high medium consumption and consequently high operating costs. A value often considered for process development is a perfusion rate of one reactor volume per day (RV/day), but depending on the specific characteristics of the system under consideration, higher perfusion rates may also be favourable. The lower bound of the perfusion rate, on the other hand, depends on multiple factors, including the cell-specific growth rate, the medium formulation and cost, and the stability of the product. Low perfusion rates lead to long product residence times in the reactor favouring post-translational modifications of the product with negative consequences on product quality attributes. For a given cell density, as the perfusion rate is decreased, metabolite levels can be reached, below which further cultivation is not possible due to induced cell death, decline in cell-specific productivity or decline in product quality (Konstantinov et al. 2006).

In the case of viable cell density, X_V, it is obvious that, in principle, larger values are always to be preferred so that only upper limits actually need to be considered. In addition to the minimum cell-specific perfusion rate, which – at a given medium throughput – constrains the maximum viable cell density, other limitations come from oxygenation, viscosity of the cell culture broth and cell retention. As mentioned before, running the cell culture below the minimum CSPR does not allow a stable operation; thus, for a given perfusion rate, the maximum VCD is constrained by the depth of the used medium, which defines the $CSPR_{min}$ value (Konstantinov et al. 2006). On the other hand, perfusion cultures are typically operated at much higher viable cell densities compared to batch cultivations, which implies larger oxygen consumption and carbon dioxide production rates. These typically become a problem when targeting cell densities above about 100×10^6 cells/mL. Another limitation related to high cell density values is that the viscosity increases, which results in higher stirring and agitation rates in order to guarantee complete homogenisation of the cell culture. This can lead to shear stress values that are harmful for the cells (Ozturk 1996). Finally, studies on filtration-based cell retention devices have shown that, depending on the specific retention technology adopted, limitations on the cell density value can occur. These limitations can be related to the residence time of the cells within the uncontrolled environment of the retention device; the characteristics of the device itself, such as the pump capacity with respect to tangential or alternating tangential flow filtration, or the porosity structure of the filtering media (Clincke et al. 2013b).

As discussed in Chapter 2, the concept of minimum cell-specific perfusion rate, $CSPR_{min}$, represents one of the most important control aspects in perfusion cultures. This is determined by the medium formulation and the expression system itself. Depending on cell line and clone, the nutrient requirements can greatly vary. Thus, early media and cell clone screening are prerequisites for process design for continuous cultures.

With respect to the expression system, for a given set of operating conditions (P, VCD, and CSPR), cellular metabolism for an ideal non-ageing cell should be fixed and characterised by constant cell-specific productivity, q_P, and cell-specific growth rate, μ (Ozturk 2014):

$$q_p = \frac{g_{protein} \quad \text{produced}}{\text{cell} \times \text{time}} \qquad \left[\frac{pg_{protein}}{\text{cell} \times \text{day}}\right] \tag{4.1}$$

$$\mu = \frac{\text{cell} \quad \text{produced}}{\text{cell} \times \text{time}} \qquad \left[\frac{1}{\text{day}}\right]. \tag{4.2}$$

These parameters are very important in defining the process performance, in terms of productivity, medium consumption and yield. Their value is defined by the cell metabolism, and therefore, it is a function of temperature, pH and composition of the supernatant, particularly in terms of nutrients and toxic metabolites.

It is worth noting that, ideally in a culture operated at constant viable cell-density, the cell specific productivity does not depend on the CSPR if the medium is deep enough (Konstantinov et al. 2006). This means that the CSPR can be decreased without changing q_P as long as no nutrient limitation is encountered or, for any other reason, the culture becomes unstable. This defines the CSPR_{min} value. On the other hand, the cell-specific growth rate, μ, decreases when decreasing the cell-specific perfusion rate, and therefore, at the CSPR_{min} the cell growth is minimum but still high enough to sustain a viable culture. These considerations highlight the relevance of the CSPR_{min} in determining the optimal operating conditions of the bioreactor.

4.1.2 The Optimal Bioreactor Design Procedure: Step 1

In order to univocally define the bioreactor steady-state operation, we need to select values for the viable cell density, the perfusion rate and, consequently, the CSPR. For this, we consider three objectives, which define the performance of a given bioreactor: the volumetric productivity (PR), the yield (Y) and the medium consumption (MC), which are defined as follows:

$$PR = \frac{g_{protein} \quad \text{produced}}{V_t \times \text{time}} = (P - B)C_P \qquad \left[\frac{g_{protein}}{L \times \text{day}}\right] \tag{4.3}$$

$$Y = \frac{(P - B)}{P} \tag{4.4}$$

$$MC = \frac{\text{media} \quad \text{consumed}}{g_{protein} \quad \text{produced}} = \frac{P}{PR} \qquad \left[\frac{L_{media}}{g_{protein}}\right], \tag{4.5}$$

where P represents the perfusion rate (RV/day), B is the bleed rate (RV/day), and C_P is the protein concentration inside the bioreactor or titre (g/L).

As seen in Section 2.6.1, the steady-state mass balances for cells and target protein lead to the following expressions for the bleed flowrate, B and the protein concentration, C_P, respectively:

$$B = \mu \tag{4.6}$$

$$C_P = \frac{q_p X_v}{P} = \frac{q_p}{\text{CSPR}}. \tag{4.7}$$

Thus, substituting in Equations 4.2–4.4, the objective functions, PR, Y and MC at steady state can be written as follows:

$$PR = (P - \mu)\frac{q_p X_V}{P} \tag{4.8}$$

$$Y = \frac{(P - \mu)}{P} \tag{4.9}$$

$$MC = \frac{P^2}{(P - \mu)\, q_p X_V}. \tag{4.10}$$

When we consider the design of a perfusion process, we refer to the previously mentioned parameters to quantify the process performance, aside from any consideration about product quality. In particular, while the volumetric productivity and yield need to be maximised, the medium consumption should be minimised. As the protein removed in the bleed stream is not further processed, the bleed is considered as product loss and therefore, according to Equation (4.9), the cellular growth, μ should be minimised to maximise the process yield, which, as mentioned before, occurs at the CSPR$_{min}$. In general, reasonable values for the bleed stream and, therefore, for the process yield are considered in the range 5–10 per cent of the overall perfusion rate. When considering productivity and medium consumption, it is generally understood that operation at CSPR$_{min}$ gives optimal process performance (Konstantinov et al. 2006). In principle, two approaches can be envisioned to estimate the CSPR$_{min}$. The push-to-low approach, originally proposed by Konstantinov et al. (2006), consists of a stepwise reduction of the perfusion rate at a fixed viable cell density so as to explore decreasing values of the CSPR. Alternatively, the push-to-high approach involves a stepwise increase of the viable cell density at a constant perfusion rate. In both cases, the CSPR decreases at each step until either medium limitations or metabolite inhibitions result in process instability (e.g., cell death or decrease in cell-specific productivity), which, in principle, occurs at the CSPR value corresponding to the CSPR$_{min}$. The path followed to reach the CSPR$_{min}$ value in the approaches is not the same. In the first one, the perfusion rate decreases, which may lead to excessive average residence times in the case of fragile products, while in the second, the VCD increases, which may lead to unfeasibly large values. The two procedures are illustrated schematically in Figure 4.1.

If we consider the operation at constant perfusion rate, the optimisation procedure comprises the following step. We take a reasonable set of starting conditions for P and X_V and measure the volumetric productivity and medium consumption at steady state. As discussed in the following paragraphs, from Equations (4.8) and (4.10), we can see that by increasing the VCD, the push-to-high approach results in increased volumetric productivity and decreased medium consumption. This is continued until the culture remains stable – that, is, until CSPR becomes equal to CSPR$_{min}$, as indicated by the point in Figure 4.1. Accordingly, at this point, for the selected perfusion rate, the volumetric productivity is maximised and the medium consumption is minimised while providing a stable cultivation. This is illustrated for the operation at a constant perfusion rate of 1.0 RV/day in Figure 4.2.

Figure 4.2A describes the relation between VCD and perfusion rate for various values of CSPR. Considering a sequential bioreactor operation at constant perfusion rate of 1.0 RV/day and viable cell densities between 10 and 50×10^6 cells/mL,

4.1 Principles of Optimal Perfusion Bioreactor Design

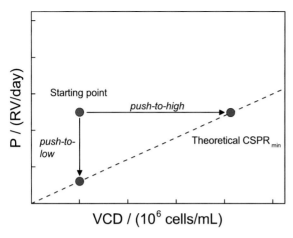

Figure 4.1 Schematic representation of the push-to-low and the push-to-high approach to determine CSPR$_{min}$. The red point represents the initial condition of the procedure and the slope of the broken straight lines the CSPR value. © Reprinted from *Biochemical Engineering Journal*, 151, Moritz K. F. Wolf, Anna Pechlaner, Veronika Lorenz, Daniel J. Karst, Jonathan Souquet, Hervé Broly and Massimo Morbidelli, 'A two-step procedure for the design of perfusion bioreactors', 2019, with permission from Elsevier

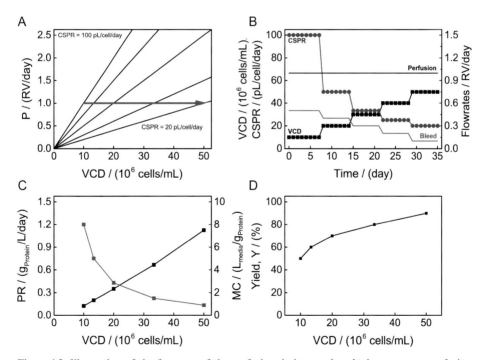

Figure 4.2 Illustration of the first step of the perfusion design: *push-to-high* at constant perfusion rate. For the given example, we consider the operation at a constant perfusion rate of 1.0 RV/day and constant cell-specific productivity of 25 pg/cell/day. Relation between perfusion rate and VCD for various values of the CSPR (A); the arrow shows the decrease of the CSPR (from 100 pL/cell/day down to 20 pL/cell/day), when increasing the VCD at constant P. VCD (■), CSPR (●), perfusion rate (black line) and bleed rate (red line) as a function of time (B). Volumetric productivity (black) and medium consumption(red) as a function of the viable cell density (C). Yield as a function of the viable cell density (D).

the resulting CSPR ranges between 100 and 20 pL/cell/day. Let us consider the case where the cell-specific productivity is assumed to be constant at 25 pg/cell/day, while the cell-specific growth rate, μ (and, therefore, the bleed rate), decreases with the CSPR. In particular, with illustrative purposes, we assume the cell-specific growth equal to 0.5 day^{-1} at a CSPR of 100 pL/cell/day and equal to 0.1 day^{-1} at 20 pL/cell/day (Figure 4.2B). Using Equation (4.7), this results in a titre increasing from 0.25 up to 1.25 g/L for decreasing CSPRs. Finally, using Equations (4.8) and (4.10), we can compute the resulting volumetric productivity, PR, and medium consumption, MC. As shown in Figure 4.2C, we see that the volumetric productivity increases linearly with the viable cell density from 0.13 up to 1.13 g/L/day, while the medium consumption decreases inversely from 8.0 to 0.89 L/g. On the other hand, we see from Equation (4.10) that since the specific cell growth rate decreases, the yield, Y, of the process also improves as the viable cell density increases, as shown in the Figure 4.2D. For this exemplary case, the yield increases from 50 per cent to 90 per cent. It can then be concluded that for the given perfusion rate P, the maximum feasible VCD constitutes the optimal operating condition, which corresponds to the value CSPR$_{min}$.

4.1.3 The Optimal Bioreactor Design Procedure: Step 2

We have seen the process optimisation for a given value of the perfusion rate, P. Now, we want to analyse how the objective functions change when we change P and X_v but keep the CSPR fixed, ideally at its optimum value CSPR$_{min}$. In the case that we express Equations (4.8) and (4.10) as a function of P and CSPR, we obtain the following relations:

$$PR = (P - \mu) \frac{q_p}{\text{CSPR}} \qquad (4.11)$$

$$MC = \frac{P \times \text{CSPR}}{(P - \mu) \, q_p}. \qquad (4.12)$$

If we do the same but now as a function of the viable cell density and the CSPR, we obtain

$$PR = \left(X_V - \frac{\mu}{\text{CSPR}} \right) q_p \qquad (4.13)$$

$$MC = \frac{X_V \times \text{CSPR}}{\left(X_V - \frac{\mu}{\text{CSPR}} \right) q_p}. \qquad (4.14)$$

If we now consider the operation at constant CSPR (and, therefore, constant q_P and μ), we see from Equations (4.9), (4.11) and (4.12) that as P increases, the productivity and the yield always increase while MC decreases so that we can conclude that the optimum perfusion rate is the maximum value that can be sustained by the set-up under consideration, P_{max}. The same is concluded by considering the viable cell density, as in Equations (4.13) and (4.14). Thus, either the maximum achievable VCD or the maximum sustainable P describe the optimum operating conditions, as illustrated in Figure 4.3 for the illustrative case of a fixed CSPR = 20 pL/cell/day. Figure 4.3A shows the relation between VCD and perfusion rate for the selected CSPR value. It is seen that operations at perfusion rates of 1–5.0 RV/day require the set-up to maintain viable cell densities between 50 and 250 $\times$ 10^6 cells/mL. (Figure 4.3B). Assuming for this system a constant specific cell growth correspond-

4.1 Principles of Optimal Perfusion Bioreactor Design

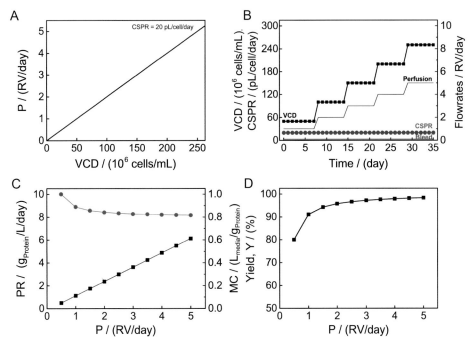

Figure 4.3 Illustration of the second step of the perfusion design procedure at constant CSPR. For the given example, we consider the operation at a constant CSPR of 20 pL/cell/day with a fixed bleed rate of 0.1 RV/day and cell-specific productivity of 25 pg/cell/day. Relation between perfusion rate and VCD for the CSPR value of 20 pL/cell/day (A). VCD (■), CSPR (●), Perfusion rate (black line), and bleed rate (red line) as a function of time (B). Volumetric productivity (black) and medium consumption (red) as a function of the perfusion rate (C). Yield as a function of the perfusion rate (D).

ing to a bleed rate of 0.1 RV/day and a constant cell-specific productivity of 25 pg/cell/day, from Equation (4.7), we compute a constant titre of 1.25 g/L. From this, using Equations (4.9), (4.11) and (4.12), we can calculate the resulting volumetric yield, Y, productivity, PR, and medium consumption, MC, as a function of P. In Figure 4.3C, we can see that the volumetric productivity is increasing linearly with the perfusion rate from 0.5 g/L/day up to 6.1 g/L/day, while the medium consumption decreases inversely from 1.0 L/g to 0.82 L/g. In particular, in the range of low perfusion rates – say, between 0.5 and 3.0 RV/day – increasing the perfusion rate has a strong effect in reducing medium consumption, which becomes less significant at higher P values. On the other hand, it is important to consider the actual feasibility of operating the bioreactor at perfusion rates above 3 RV/day, but especially at viable cell densities above 100×10^6 cells/mL (equal to 2 RV/day at a CSPR of 20 pL/cell/day). In Figure 4.3D, it is seen that the calculated yield increases from 80 per cent to 96 per cent by a step-by-step increase of the perfusion rate from 0.5 RV/day to 2.0 RV/day, and then continues to increase but much slower up to 98 per cent as the perfusion rate increases further. This shows the importance of the optimisation procedure at constant CSPR, as there is a large optimisation potential, but also feasibility constraints that have to be taken into account. The value of the maximum feasible perfusion rate, $P_{\max}$ depends on the expression system, the chosen medium and the reactor.

From the preceding discussion, we see that the design of a perfusion process requires a sequence of two experimental steps: in the first one, we evaluate the $CSPR_{min}$, and in the second, the maximum sustainable perfusion rate, P_{max} and viable cell density, $X_{V,max}$ (with ratio equal to $CSPR_{min}$), are determined.

4.2 DEVELOPMENT OF A COMMERCIAL PERFUSION CULTURE

The development of a commercial perfusion process for a new cell culture requires a sequence of steps involving experiments ranging from the μL to the L-scale of the laboratory bioreactor and finally to the thousands of litres of the commercial scale. We now introduce the basic concepts of the procedure and next discuss a series of case studies.

The procedure for developing a new commercial perfusion culture is illustrated in Figure 4.4 as constituted of three parts: clone and media screening, perfusion cell culture development and process scale-up. Each one of these parts has its own objectives and requires different technologies, which are discussed separately in the following paragraphs.

Typically, at the beginning of the process development, we have a few hundred clones, coming from the cell line generation and development, that can be combined with several medium compositions for the expression of a given target protein (Part 1). In this first step, the objective is to screen the two or three most promising clones and one or two basic medium compositions in order to proceed with the next step of process development. As objective parameters for the selection, we consider the cell-specific productivity and growth rate, with no reference to product quality or the fact that the culture may be stable or not in perfusion. These two quantities have to be measured for a large number of combinations clone/medium. This requires high-throughput techniques and experiments at the very small scale (μL–mL range).

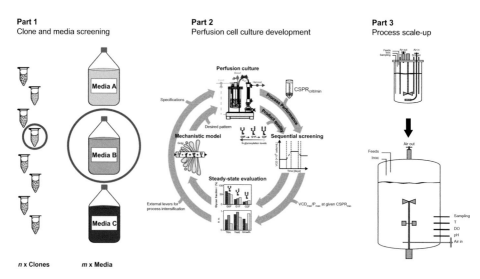

Figure 4.4 Summary of the commercial perfusion culture development procedure: Part 1: Clone and media screening. Part 2: Perfusion cell culture development. Part 3: Process scale-up.

The most common approach is to use scale-down models and particularly high-throughput devices, such as deepwell plate systems. As discussed in Chapter 3, these devices are usually designed for batch or fed-batch operating mode and, therefore, are not best suited for mimicking perfusion operations. Accordingly, in the following, we consider semi-continuous high-throughput units.

Using the few clones and media identified in the first step, we now proceed to identify a set of stable operating conditions for a perfusion bioreactor with optimal performance in terms of maximum productivity and minimum medium consumption. The final result is the determination of the most convenient viable cell density and perfusion rate (Part 2). It has to be mentioned that the product yield does not need to be considered explicitly, since the bleed rate is fixed by the specific cell growth rate of the selected cell line (Equation (4.6)), and therefore, higher productivities imply better yields (Equation (4.9)). Performing all the experiments necessary for this task in the litre scale is too slow and expensive, and therefore, we need again suitable scale-down models that mimic perfusion-like experiments.

In the case that the obtained reactor performance is not satisfactory, it is necessary to go back and to reconsider some of the already selected operating conditions, such as the operating temperature or some chemical additives that may significantly alter the system and then require a new design of the reactor operating conditions. The same situation occurs in the case where the obtained operating conditions do not lead to satisfactory product quality parameters. For example, in the case that the N-linked glycosylation pattern of the product needs to be modified, we have to add some proper feed supplements and then repeat the reactor design procedure.

The objective of the third part is to transfer the perfusion process, developed at the benchtop bioreactor scale, to the larger clinical and commercial scales (Part 3). This means to change the size and shape of the reactor (if needed) and the ancillary equipment, as well as the corresponding operating conditions, in order to reproduce the process performance and product quality attributes obtained at the laboratory scale. At this step, no attempt is made to further improve any of the process or product parameters; the idea is that satisfactory performances have been obtained at the laboratory scale, and now they have to be replicated at larger scales. For this, we have to, for example, make sure that mixing is sufficient, mass transfer does not become limiting and cells do not experience too high shear stress. This is a typical task in chemical engineering, and a whole set of tools has been developed for this in the general area of multiphase reactors, as discussed in Chapter 5.

4.3 PART 1: CLONE AND MEDIA SCREENING FOR PERFUSION PROCESSES

As discussed in Chapter 1, cell line development starts with the transfection of the expression vector to the host cell line. This includes the random integration of the foreign genes of interest, which leads to a heterogeneous pool of clones with different growth and protein productivity characteristics. Typically, clones with high and stable recombinant protein production are preferred. However, in a population of transfected cells, only a few clones are high producing, as the protein expression relies on the full integration of the genes for the heavy- and light-chain antibody fragments. As this full integration of the active region supporting high gene expression in the chromosomes of the host cell line is rare (Browne & Al-Rubeai 2007), the selection of proper clones represents a critical and particularly laborious task. In addition, such high-producer cell clones typically exhibit lower growth rates, as most resources are used for protein expression. Therefore, screening of a very large number of clones is necessary in order to have a good chance to find a few high-producing clones with good cellular growth characteristics (Lai et al. 2013). This step requires high-throughput platforms in order to test and screen large numbers of clones and media at the same time. As mentioned before, the selection criteria are solely based on growth rate and productivity.

Historically, this step is performed in batch mode for both fed-batch and perfusion applications. Typically, cells are sequentially diluted on well plates to screen for high-producer cell clones, thus mimicking a batch-operating mode. However, clones and media that are best for fed-batch may not be optimal for perfusion. For example, a fast growth rate is desired for fed-batch processes in order to accumulate a large number of cells in a short time during inoculation so that these high-producer clones produce as much protein as possible during the production phase. On the other hand, a high growth rate leads to high bleed rates (Equation (4.6)) in order to maintain a constant viable cell density during the steady-state production phase of a perfusion bioreactor and, thus, as shown in Equation (4.9), to low product yields or high product losses. In addition, a very high growth rate triggers high generation numbers and, thus, can lead to cellular instability in perfusion applications operated at very long run durations (Ozturk 2014). Thus, for perfusion cultures, a stable cellular growth rate and high cell-specific productivity are more important than targeting the highest end titre and cell density. In conclusion, we need high-throughput screening devices that can mimic perfusion-like processes and enable long-time screening in order to test the long-term stability and high productivity of the clones of interest.

4.3.1 High-Throughput Systems

Several requirements need to be fulfilled in order to perform cell clone and media screening for a perfusion process. The most critical one is probably to mimic the perfusion operating mode. We can distinguish between two different approaches: non-steady-state and steady-state operation. In non-steady state-perfusion processes, as mentioned in Chapter 2, there is no bleed stream to remove the cells, which, therefore, continuously accumulate in the reactor with respect to their specific growth rate. Accordingly, only medium addition and cell-free harvest are present in this approach. Steady-state perfusion processes, on the other hand, require the cell-removing bleed stream in order to operate at a constant viable cell density, stable in time.

Next we need a scale-down high-throughput system. This should be operated in the nL, μL scale or low mL scale and should allow parallel operation of several combinations of clones and media at the same time. For this, a high degree of automation is needed in order to reduce manual work. Potential candidate systems for this application are the custom nanofluidic chips using Berkeley Lights OptoElectro Positioning technology, deepwell plates (DWP), shake-tube (ST) bioreactors and the ambr$^{©}$ 15 bioreactors, which have been introduced and described in Chapter 3. In order to illustrate the use of such scale-down models for clone screening, we discuss in the following an example using deepwell plates and shake-tube bioreactors and illustrate the validation of the obtained results at the bench scale.

In particular, we consider a semi-continuous procedure, referred to as the VCD_{max} experiment, that allows the operation of perfusion-like processes at constant medium exchange rate in shake tubes and deepwell plates (Bielser et al. 2019b). The procedure consists of a daily viable cell count, the separation of medium and cells, removal of cell-free supernatant and replacement by fresh medium, without additional cell removal. By repeating this procedure on a daily basis, the cell concentration increases in time until a maximum value – that is, the maximum viable cell density (VCD_{max}) sustainable at the given perfusion rate. This procedure has been shown to work in shake-tube bioreactors in a variety of applications (Gomez et al. 2017, Villiger-Oberbek et al. 2015), and also in deepwell plates (Lin et al. 2017). Bielser et al. (2019b) validated the procedure in shake tubes (15 mL working volume) and 96-DWP (450 µL working volume) and applied it to screen a set of 12 different CHO cell clones expressing the same IgG1 antibody. The results were compared in terms of VCD_{max} and volumetric productivities.

In the shake tubes, a centrifugation step was applied to separate cells and medium, and the medium in the entire reactor volume was replaced once per day. In the 96-DWP, liquid handling was performed through a robotic platform (Biomek FX, Beckman-Coulter, Fullerton, CA, USA). Here, cells were left to settle for one hour in static conditions in order to separate cells from the medium before half a reactor volume was exchanged. Removing the entire reactor volume was not done due to difficulties in controlling the height of the tip which could lead to additional unwanted cell removal. Cell settling, moreover, results in a less dense cell pellet compared to the centrifugation step applied for the shake-tube system. The results obtained in the ST and the 96-DWP systems were compared, in a first step, in terms of the volumetric productivity at VCD_{max}. The different clones were ranked based on volumetric productivity measured in the two systems. Three groups were identified in each system: high producer (two clones), medium producer (three clones), and low producer (seven clones). The different clones did not share the exact same ranking in the two systems, but at least they were in the same groups, as illustrated in Table 4.1. In a next step, four of these clones were selected to validate the results in benchtop bioreactors at lab-scale (3.5 L working volume).

The clone selection for the benchtop operation was made so as to have sufficient diversity among the different clones: one clone was a higher grower (high VCD_{max}) and gave a high mAb titre in the semi-continuous operation and one clone exhibited low growth (low VCD_{max}) and low mAb titre while the third clone was a high grower but a low producer, and the fourth one was a low grower and a high producer. The results obtained in the two scale-down systems could be validated at lab scale, which, in fact, gave $CSPR_{min}$ and productivity values at steady state similar to those obtained with the VCD_{max} experiments.

Thus summarising, this study demonstrates the potential of the 96-DWP and the shake-tube systems for screening a variety of different clones. In addition, the semi-continuous operation nature of these systems enabled a first prediction of the minimum cell-specific perfusion rate for the considered expression systems. For two of the employed cell clones, the predicted volumetric productivity and cell-specific perfusion rate in the scale-down systems were subsequently validated through long-term benchtop bioreactor runs. The two scale-down systems appear to be very useful for the screening of different media and cell line performances in perfusion mode. The design of the VCD_{max} experiment indeed allows one to make a first estimate of the minimum CSPR but on the other hand, does not allow one to test the cell

Table 4.1 Ranking of clones 1–12 based on the volumetric productivity values computed from the corresponding mAb concentrations reached at VCD_{max} in ST and 96-DWP, Clones 1 and 5 are highlighted in bold because they were identified as the two top producers in both screening methods. The four clones selected for further investigation in bioreactors were marked with symbol (♦). © Reprinted from *Biotechnology Progress*, Jean-Marc Bielser, Jakub Domaradzki, Jonathan Souquet, Hervé Broly and Massimo Morbidelli, 'Semi-continuous scale-down models for clone and operating parameter screening in perfusion bioreactors', 2019, with permission from John Wiley and Sons.

Clone #	Volumetric Productivity in ST (g/L_R/day)	Volumetric Productivity in 96-DWP (g/L_R/day)	Rank in ST	Rank in 96-DWP
#1♦	**1.14**	**0.59**	**1**	**2**
#5	**1.01**	**0.67**	**2**	**1**
#6	0.85	0.51	3	4
#7	0.82	0.47	4	5
#2 ♦	0.67	0.53	5	3
#4 ♦	0.51	0.28	6	9
#8	0.49	0.18	7	11
#9	0.46	0.45	8	6
#3 ♦	0.45	0.36	9	7
#10	0.43	0.25	10	10
#11	0.36	0.10	11	12
#12	0.27	0.33	12	8

specific long-term stability. In terms of ease of operation and medium consumption, the DWP system is better than the ST system, but it would still be outperformed by the nanofluidic chip technology presented in Chapter 3.

4.3.2 Considerations on Screening Operation

Clone and medium screening requires, indeed, very small working volumes and needs to take place at the single cell level. We described a variety of scale-down models in Chapter 3, and their selection should be based on the specific objective of the planned experiment and the corresponding stage in process development.

At the very early stage, with hundreds of clones involved, the traditional plate-based workflows and especially the nanofluidic chip device offer the most attractive solutions. As mentioned before, a screening in perfusion mode, which seems to be feasible in the Beacon platform, would be desirable to identify promising candidates for perfusion early. This is definitely done in a second stage, after reducing the number of clones to tens of clones, using the 96-DWP system and the VCD_{max} experiments, which mimic the perfusion operating mode. In this context, it could be of interest to evaluate the 384-DWP systems for such experiments.

The long-term stability testing probably needs to be performed at larger scale, using either the shake-tube system or the ambr© 15 system applying sedimentation, or even at larger scale like the ambr© 250 perfusion system (Chotteau 2017, Zoro & Tait 2017). The suitability of the different systems for the different stages of cell line and media screening is summarised in Table 4.2.

Table 4.2 Comparison of various high-throughput systems with respect to their suitability for the different stages of the cell line and media screening processes. A plus sign indicates the system is promising, a minus sign indicates the system is not suited for the specific task.

Application	Beacon platform	96- / 384-DWP	Shake tubes (10–20 mL)	ambr15	250 mL bioreactors
Cell line and media screening	+	+	−	−	−
Suitability for perfusion	+	+	+	+	+
Long-term stability	+	−	+	+	+

4.4 PART 2: PERFUSION CELL CULTURE DEVELOPMENT

In this step, we identify, for the few selected clone/medium combinations, the best operating conditions for a stable perfusion process at the litre scale targeting maximum productivity ($g_{protein}$/L/day) and minimum medium consumption (L_{medium}/$g_{protein}$). Already at the litre scale, perfusion processes require large amounts of medium. Operating a perfusion process in a 2 L reactor at a perfusion rate of 1 RV/day for 30 days requires 60 L of perfusion medium. If one considers the number of experiments that have to be performed in order to determine optimum operating conditions, it appears that the whole procedure would require a long time and large medium volumes. Therefore, we need to use scale-down models that allow operation at the µL or mL scale.

The general procedure for the optimal design of a perfusion bioreactor, in terms of perfusion rate and viable cell density, has been described in Section 4.2 based on the two steps illustrated in Figure 4.5: (1) determination of the minimum cell-specific perfusion rate and (2) the determination of the best P (or VCD) for the given $CSPR_{min}$. The $CSPR_{min}$ describes the minimum possible value of the CSPR to have a sustainable perfusion culture. Therefore, for a given expression system (medium and clone), at $CSPR_{min}$ the cellular growth cannot be further decreased and cell-specific productivity cannot be further increased without changing media or clones.

The reactor performance obtained at this point may still not be satisfactory and, therefore, requires some further optimisation steps. We consider, as examples, the two further steps illustrated in Figure 4.5. The first one refers to the case, where the specific cellular growth is still high at the identified $CSPR_{min}$ and therefore leads to excessive product losses in the bleed. In order to improve the process yield, it is then possible to introduce changes in the operating conditions to inhibit cell growth (Lin et al. 2017, Wang et al. 2018, Wolf et al. 2019a).

The fourth step refers to product quality. This has to satisfy certain predefined specifications and in some cases, such as for biosimilars, it has to be sufficiently similar to a desired quality pattern. When this is not the case, we need to take action, which is mostly done through the introduction of proper nutrient additives. In Section 4.5.3 and in particular in Chapter 6, we discuss the combination of mechanistic modelling and the experimental investigations to modulate certain product quality attributes through proper continuous feed additions (Karst et al. 2017b; Villiger et al. 2016a, 2016b).

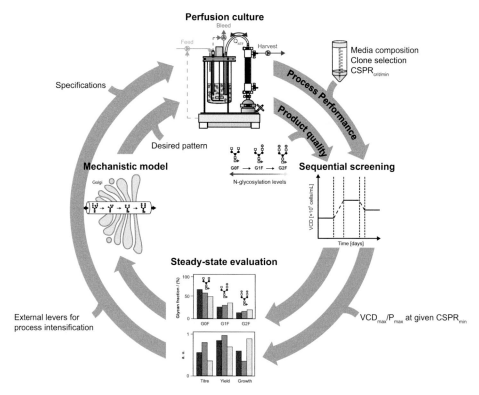

Figure 4.5 Schematic of the procedure for the optimal bioreactor design. In the first step, the minimum CSPR is identified, and in the second, the maximum volumetric productivity at the $CSPR_{min}$ is defined. In the third step, the process yield is optimised through cell-growth inhibition, and in the last, product quality is modulated to match a desired pattern. The procedure is iterated until converging to satisfying overall reactor performance.

At the end of both the third and the fourth steps, the procedure for the design of the perfusion rate and the viable cell density has to be repeated, thus leading to the iterative procedure illustrated in Figure 2.5, which ends when all aspects of the perfusion culture are considered satisfactory. The scope of the following sub-sections is to provide examples of application of this procedure, involving in all cases systems of commercial interest.

4.4.1 Definition of Minimum CSPR

Being the first step of the procedure, this is certainly the most challenging one, since we still have very little knowledge about the expression system. Minimising the CSPR can be done by applying the push-to-low/push-to-high approach shown in Figure 4.1. This is done in one experiment where, starting from a steady-state at a relatively high CSPR value, the bioreactor is forced to go through a sequence of steady states characterised by decreasing CSPR values. As described in Figure 4.1, this can be done by either step wise decreasing P, while keeping VCD constant, or decreasing stepwise VCD with P unchanged. These experiments at the litre scale require very long times and the consumption of relevant amounts of media, thus calling for scale-down models in the mL or μL scale.

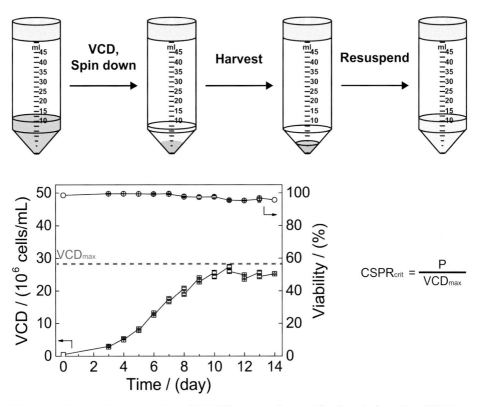

Figure 4.6 Schematic representation of the VCD_{max} experiment. After inoculation at low VCD (e.g. 0.5×10^6 cells/mL), STs are operated in a semi-continuous way. Every 24 h, cells are counted, spun down, the cell-free supernatant harvested, replaced by fresh medium and the cells resuspended. This operation results in the maximum viable cell density achievable at the given medium exchange rate, and allows determining the critical value $CSPR_{crit}$. © Reprinted from *Biotechnology and Bioengineering*, Moritz K. F. Wolf, Andrea Müller, Jonathan Souquet, Hervé Broly and Massimo Morbidelli, 'Process design and development of a mammalian cell perfusion culture in shake-tube and benchtop bioreactors', 2019, with permission from John Wiley and Sons

VCD_{max} Experiments: Evaluation of $CSPR_{crit}$. The VCD_{max} experiments described in Section 4.4.1 and used for clone/media screening in DWP (µL) and ST (mL) units provide the first step for the procedure to define $CSPR_{min}$. The procedure consists of a daily viable cell count, the separation of medium and cells, removal of cell-free supernatant and medium replacement by fresh medium, before either shake tubes or deepwell plates are transferred back to the incubator without having removed any cell, as illustrated in Figure 4.6. By repeating this procedure on a daily basis, the maximum viable cell density (VCD_{max}) achievable at the given perfusion rate can be evaluated. This corresponds to the viable cell density value where the provided amount of nutrients per cell per day (CSPR) becomes critical and cellular density stops increasing or even starts decreasing, as shown in Figure 4.6. Note that this type of experiment does not include cell removal (bleeding) and, therefore, cannot assess the long-term stability of the perfusion run. Accordingly, we refer to this as the critical CSPR, $CSPR_{crit}$, which should not be confused with the $CSPR_{min}$.

In the case of Bielser et al. (2019b), the evaluated $CSPR_{crit}$ in DWP and ST was found to lead to long-term stable perfusion runs in the bioreactor, and therefore, it was not distinguished from the $CSPR_{min}$. However, in the case of Wolf et al. (2019b), the viable cell density values measured in these experiments – i.e., $VCD_{max} = 27.5 \times 10^6$ cells/mL – could not be maintained in the long-term bioreactor operation, for which a lower value – i.e., $VCD = 20 \times 10^6$ cells/mL – had to be used. Such a different behaviour can indeed be attributed to different cell lines used in the two experiments. In all cases, it is recommended not to rely only on experiments at this small scale for the definition of the $CSPR_{min}$ but to validate it at a larger scale with long-term stability experiments at constant viable cell density, thus equipment allowing for cell removal.

It has to be noted that, if the identified VCD_{max} corresponds to the complete depletion of one or more nutrients in the used medium, a medium with higher depth can be prepared, and the procedure can be repeated, as discussed for the push-to-low and push-to-high steady-state experiments in Konstantinov et al. (2006). It should be considered that this may lead to higher osmolality values in the reactor, which needs to be accounted for.

Steady-State Experiments in the mL Scale. Once the VCD_{max} and $CSPR_{crit}$ values have been obtained from the perfusion scale-down models discussed earlier, we need to proceed to identify the $CSPR_{min}$ – that is, to identify the minimum sustainable CSPR at which a stable culture can be operated for longer times. As mentioned before, such a $CSPR_{min}$ value can be either equal to or larger than the $CSPR_{crit}$, which therefore provides an excellent starting point. This can be done either directly in the bioreactor at the litre scale or, even better, in a similar scale-down model.

Wolf et al. (2018) developed a shake-tube-based scale-down model in order to investigate steady-state operations in the mL scale, referred to in the following as VCD_{ss}. The developed procedure is similar to the one for the VCD_{max} experiment but with two significant differences. Firstly, the shake tubes are inoculated at very high cell densities close to the targeted viable cell density. Secondly, in order to operate at a constant viable cell density, a bleed step is introduced. This comes in addition to the cell removal connected with the culture sampling for the cell count, and it is specifically designed to keep the VCD constant around a predefined target value ($X_{V,target}$), as illustrated in Figure 4.7. The bleed and harvest rates are mimicked by removing cell-containing and cell-free volumes, respectively, from the ST, which is calculated on a daily basis after the cell count. The first one has to be computed so that the VCD value remains close to $X_{V,target}$ through all the cycles. In particular, we have to remove a certain cell-containing volume in order to decrease the VCD (after fresh medium addition) to a value, $X_{V,SP}$, such that the average between $X_{V,SP}$, and the measured VCD, $X_{V,meas}$), is equal to $X_{V,target}$. Accordingly, based on the measured VCD, $X_{V,meas}$, the bleed volume can be calculated as follows:

$$V_{Bleed} = \frac{(X_{V,meas} - X_{V,SP}) V_{Tot}}{X_{V,meas}}, \qquad (4.15)$$

where V_{Tot} represents the total working volume in the ST and V_{Bleed} is the total cell-containing volume to be removed from the ST including the volume of the sample taken for the cell count. The harvest volume is consequently calculated as follows:

$$V_{Harvest} = V_{Exchange} - V_{Bleed}, \qquad (4.16)$$

where $V_{Exchange}$ represents the targeted volume that should be replaced with fresh medium to simulate a given perfusion rate. For example, in the case of perfusion rates

4.4 Part 2: Perfusion Cell Culture Development

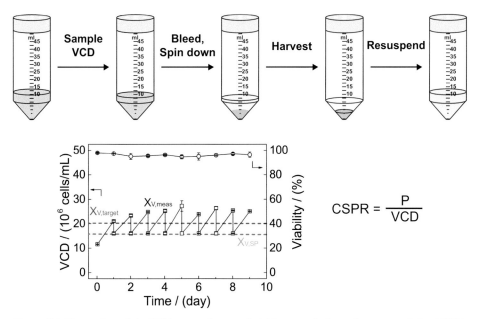

Figure 4.7 Illustration of the VCD$_{ss}$ experiment mimicking a perfusion culture at predefined VCD, $X_{V,target}$ and constant perfusion rate, P. Cells are inoculated close to the VCD bleeding set point ($X_{V,SP}$), then STs are operated in a semi continuous way. Every 24 hours, cells are withdrawn and counted. Some cells are removed based on the measurement and the remaining cells are spun down. The remaining cell-free supernatant is harvested, and the cell pellet is resuspended in fresh medium. The plot shows the VCD value oscillating above and below the predefined value, $X_{V,target}$, at the given medium exchange in the case of a stable operation – i.e., CSPR larger or equal CSPR$_{min}$. © Reprinted from *Biotechnology and Bioengineering*, Moritz K. F. Wolf, Andrea Müller, Jonathan Souquet, Hervé Broly and Massimo Morbidelli, 'Process design and development of a mammalian cell perfusion culture in shake-tube and benchtop bioreactors', 2019, with permission from John Wiley and Sons

of one reactor volume per day, $V_{Exchange}$ is equal to one reactor volume, that is, to V_{Tot}. The flowrates in the corresponding continuous perfusion are then calculated as follows:

$$B = \frac{V_{Bleed}}{V_{Tot}} \frac{1}{\Delta t} \qquad (4.17)$$

$$H = \frac{V_{Harvest}}{V_{Tot}} \frac{1}{\Delta t}. \qquad (4.18)$$

This scale-down model has been compared with a benchtop bioreactor operated at similar VCDs and perfusion rates (Figure 4.8A). It is shown that, not only are the bleed and harvest rates (Figure 4.8B) in the two systems very similar, but the entire culture develops along similar lines. This includes metabolite consumption or production rates, such as glucose and ammonia, cell-specific growth rates, titre and even the charge variant pattern of the target product (Figure 4.8C, D and F). On the other end, it is worth mentioning that the ST model differed in the lactate production and the product glycosylation pattern observed in the bioreactor (Figure 4.8C and E). This may be because of both the decreased removal of CO_2 and the lower dissolved oxygen levels due to the smaller gas–liquid mass transfer rate observed

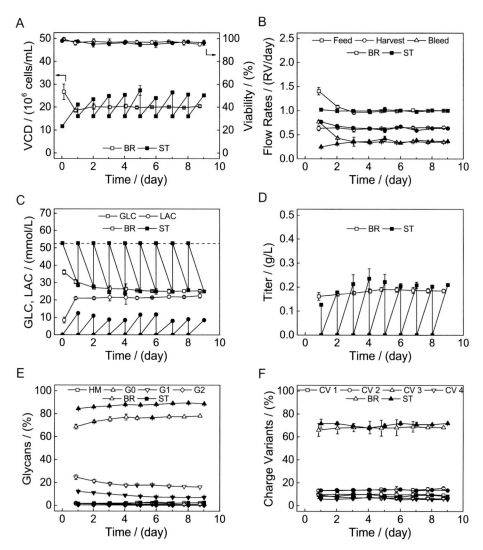

Figure 4.8 Comparison of a steady-state perfusion cell culture performed in benchtop BR and semi-continuous shake-tube (ST) bioreactors at a VCD set point of 20×10^6 cells/mL. VCD and viability (A), process exchange flowrates (B), glucose and lactate concentrations (C), product titre (D), N-linked glycoforms (E) and charge variants (F). © Reprinted from *Biotechnology and Bioengineering*, Moritz K. F. Wolf, Veronika Lorenz, Daniel J. Karst, Jonathan Souquet, Hervé Broly and Massimo Morbidelli, 'Development of a shake-tube based scale-down model for perfusion cultures', 2018, with permission from John Wiley and Sons

in the ST. Here this transfer occurs through the filter in the cap of the shake tube and is therefore less efficient than the gas spargers used in benchtop bioreactors. This results in more acidic conditions, to which the cells counteract by producing less lactate to maintain the pH at acceptable levels. Concerning the glycan distribution, the less controlled oxygen environment and the difference in oxygenation are expected to result in lower dissolved oxygen concentration in the ST compared to the benchtop system. Lower DO levels are reported to affect the mAb glycosylation, such that the level of galactosylated glycans decreases (Kunkel et al. 2000; 1998). Besides this aspect, the developed procedure enabled the mimicking of steady-state perfusion processes in bioreactor at the benchtop scale and can therefore be used for designing their operating conditions, at least in terms of perfusion rate and viable cell density.

In conclusion, it is convenient to combine the VCD_{max} experiment described earlier with the VCD_{ss} procedure, in order to first estimate a suitable VCD_{max} and the corresponding $CSPR_{crit}$, which is then taken as a starting point to determine the $CSPR_{min}$. This approach is very advantageous in practice since the simple VCD_{max} experiment easily provides a first impression about the behaviour of the selected expression system (clones and media) at a given perfusion rate and is, therefore particularly, suitable for screening applications. This experiment results in a first guess of the $CSPR_{min}$, referred to as the $CSPR_{crit}$, which needs to be further elaborated to obtain the $CSPR_{min}$ with respect to long-term stability using, for example, a benchtop bioreactor. On the other hand, the VCD_{ss} experiments can be used for the same objective of defining the $CSPR_{min}$, also considering that they provide reliable estimates of various process parameters in the bioreactor, such as the VCD and the harvest and bleed flowrates at a given perfusion rate, as well as cell-specific metabolism and a first prediction of product quality patterns. As these experiments can be performed in DWP (VCD_{max}) and ST (VCD_{max} and VCD_{ss}), they need only very small amounts of media and can be performed in parallel, which therefore results in significant cost and time savings. It can be expected that even better scale-down models could be obtained by realising more automated and better controlled systems that allow continuous media exchange at a small scale, in the direction of the work done by Chotteau (2017) and Zoro & Tait (2017) at the 250 mL scale. Alternatively, it is also possible to perform the push-to-low and push-to-high experiments in lab-scale bioreactors (1–10 L), as illustrated by Konstantinov et al. (2006) and Wolf et al. (2019c). In such a case, a sequential bioreactor run is performed, as illustrated in Figure 4.2B. However, this operation requires much higher media volumes compared to scale-down models.

4.4.2 Optimisation of Volumetric Productivity at $CSPR_{min}$

After determining the $CSPR_{min}$, the process can be further improved in terms of volumetric productivity and medium consumption, as discussed in Section 4.1.2. This can be done in either ST or benchtop bioreactors with increasing P or VCD while keeping constant $CSPR = CSPR_{min}$. By operating at constant CSPR, the cellular activity in terms of q_p and μ remains constant (Ozturk 2014), and then, as shown by Equations (4.9), (4.11) and (4.12), the volumetric productivity and the yield increase and the medium consumption decreases, as the perfusion rate increases. In addition, the average product residence time decreases, leading to a general improvement of the product-related quality attributes. Thus, overall, as the perfusion rate increases, the bioreactor performance improves in all aspects. This, however, cannot be continued indefinitely since, at some point, some limitation is encountered because of mixing intensity, oxygen supply, carbon dioxide removal or something else.

This has been illustrated by Wolf et al. (2018) in a series of VCD_{ss} experiments at 20 and 40×10^6 cells/mL with perfusion rate of 1 and 2 RV/day, respectively, so as to have the same value of CSPR of 50 pL/cell/day. To match a perfusion rate of

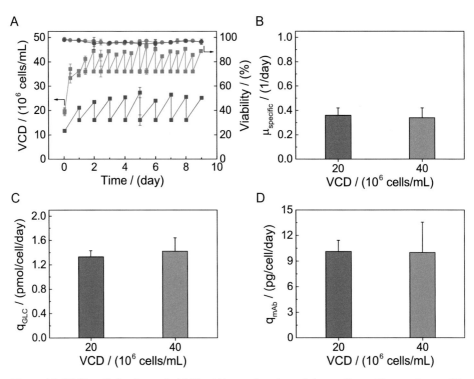

Figure 4.9 Viable cell density and viability (A) as a function of time, cell specific growth rate (B), glucose consumption rate (C) and cell-specific mAb production rate (D) as a function of the viable cell density for two experiments at the same CSPR of 50 pL/cell/day in a ST system. © Reprinted from *Biotechnology and Bioengineering*, Moritz K. F. Wolf, Veronika Lorenz, Daniel J. Karst, Jonathan Souquet, Hervé Broly and Massimo Morbidelli, 'Development of a shake-tube based scale down model for perfusion cultures', 2018, with permission from John Wiley and Sons

2 RV/day in the ST system, two full media exchanges per day – that is, every 12 instead of 24 h – was performed. As expected, similar cell-specific growth rates, glucose consumption rates and specific antibody production rates were found, as shown in Figure 4.9. On the other hand, the process performance improved in terms of productivity, medium consumption and yield, as shown in Figure 4.10.

As the ST system is quite limited, especially in terms of oxygen supply and carbon dioxide removal, it is generally not possible to operate steady-state experiments at viable cell densities above 50×10^6 cells/mL (Villiger-Oberbek et al. 2015, Wolf et al. 2018). Thus, we can follow an alternative approach and perform this step of the design procedure in a benchtop-reactor system.

The preceding results were confirmed using a benchtop bioreactor, operated with the same expression system and again at the same CSPR = 50 pL/cell/day, as shown in Figure 4.11 (Wolf et al. 2019c). A sequence of three steady-state experiments were conducted at VCD values of 20, 40 and 30×10^6 cells/mL (Figure 4.11A), with perfusion rates of 1, 2 and 1.5 RV/day, respectively (Figure 4.10B), in order to maintain the constant CSPR value (Figure 4.11C). As a result, the cells exhibited constant growth rate resulting in a constant bleed rate (Figure 4.11B) and constant metabolite environment (Figure 4.11D and E). The same behaviour is exhibited by the glycoform distribution shown in Figure 4.11F as a function of time, which remains substantially

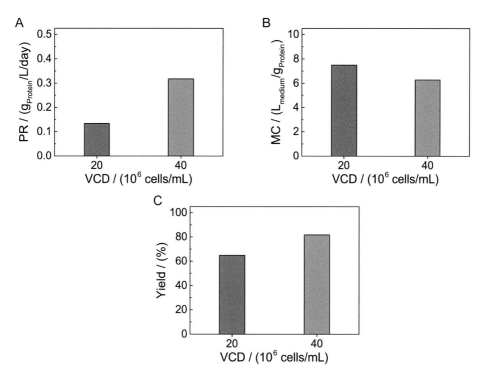

Figure 4.10 Volumetric productivity (A), medium consumption (B) and harvest yield (C) as a function of the viable cell density for two experiments at the same CSPR of 50 pL/cell/day in a ST system.

constant across the three steady-states. Based on these results and Equation (4.7), one should also expect a constant antibody concentration, which, however, as seen in Figure 4.12, was not the case.

This anomaly is due to an intrinsic instability of the used cell clone, which has been consistently observed in various experiments with the same cell line (Karst et al. 2016, Steinebach et al. 2017, Wolf et al. 2019c), which leads to the decrease of the cell-specific productivity, q_p, in time shown in Figure 4.12.

Unfortunately, this masked the effect of perfusion rate on productivity and media consumption, as apparent in Equations (4.11) and (4.12), respectively. However, the specific growth rate was apparently not affected by this cell intrinsic instability and therefore, according to Equation (4.9), the effect of P on the process yield, Y could still be observed correctly. If fact, as shown in Figure 4.11B, the bleed rate remained pretty constant during the experiment as the perfusion rate increased from 1.0 to 1.5 and then to 2.0 RV/day. This resulted in a stepwise increase of the yield from 62 per cent to 85 per cent, as shown in Figure 4.13.

The examples discussed so far illustrated the procedure for the optimal bioreactor design developed in the two steps in Section 4.2. We now proceed with the further development of the culture process in the case where the performance obtained at this point was not satisfactory for some reason, such as low yield or large media consumption or others. A good example is the case seen earlier, where the selected expression system exhibits undesired long-term instability or ageing. Some methods to further improve the process performance are discussed in the next section.

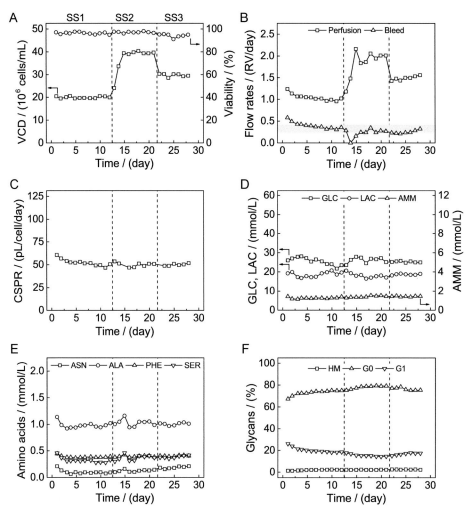

Figure 4.11 VCD and viability (A), perfusion and bleed rate (B), CSPR (C), GLC, LAC, AMM concentrations (D), amino acid concentrations (E) and glycan distribution (F) as a function of time in a sequence of three steady-state experiments at constant CSPR = 50 pL/cell/day. © Reprinted from *Biochemical Engineering Journal*, 151, Moritz K. F. Wolf, Anna Pechlaner, Veronika Lorenz, Daniel J. Karst, Jonathan Souquet, Hervé Broly and Massimo Morbidelli, 'A two-step procedure for the design of perfusion bioreactors', 2019, with permission from Elsevier

4.4.3 Further Actions for Improving Process Performance

As mentioned before, clones best suited for fed-batch culture should exhibit high cell-specific growth rates and, in particular during the stationary phase, high cell-specific productivities. On the other hand, considering a steady-state perfusion process at $CSPR_{min}$, cells should exhibit low cell-specific growth, thus resulting in high process yields and high and stable cell-specific productivities. As perfusion processes last longer than fed-batch processes, more consistent and stable cellular activities are required. Accordingly, in several studies, researchers have investigated the possibility to improve the performance of a fed-batch-platform expression system when

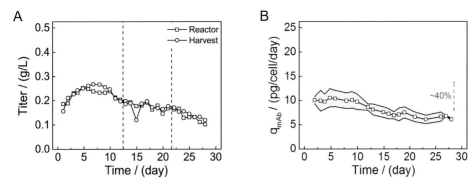

Figure 4.12 Titre in reactor and harvest (A) as a function of time in the sequence of three steady-state experiments at constant CSPR = 50 pL/cell/day in Figure 4.10, and historically observed decline in the cell specific productivity observed in these and other experiments. © Reprinted from *Biochemical Engineering Journal*, 151, Moritz K. F. Wolf, Anna Pechlaner, Veronika Lorenz, Daniel J. Karst, Jonathan Souquet, Hervé Broly and Massimo Morbidelli, 'A two-step procedure for the design of perfusion bioreactors', 2019, with permission from Elsevier

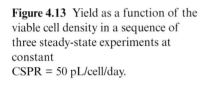
Figure 4.13 Yield as a function of the viable cell density in a sequence of three steady-state experiments at constant CSPR = 50 pL/cell/day.

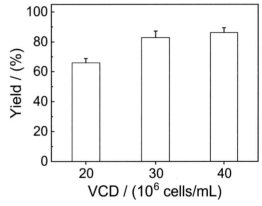

used in a perfusion bioreactor. These include the implementation of chemical and environmental growth inhibition and rebalancing of the medium composition based on the investigated energy metabolism of the cells (Du et al. 2015, Lin et al. 2017, Wang et al. 2018, Wolf et al. 2019a). These studies target especially active growth inhibition as soon as the production phase is reached, in order to reduce cellular growth and trigger target protein production.

In this frame, Wolf et al. (2019a) considered the application of induced cell-cycle inhibition for cellular growth control. The cell cycle comprises four different phases: G0/G1, S, G2 and M (Pardee 1989, Watanabe et al. 2002). The progression between the different phases is controlled by specific complexes of cyclins and cyclin-dependent kinases. Arresting the cell cycle in the G0/G1 phase would be particularly convenient since, in addition to inhibiting cell doubling, this would also maximise the protein production metabolism of the cell (Barberis et al. 2007, Sherr & Roberts 1999). For this, Wolf et al. (2019a) tested valeric acid, a short-chain fatty acid reported to be a promising G0/G1 inhibitor and productivity enhancer, as a chemical

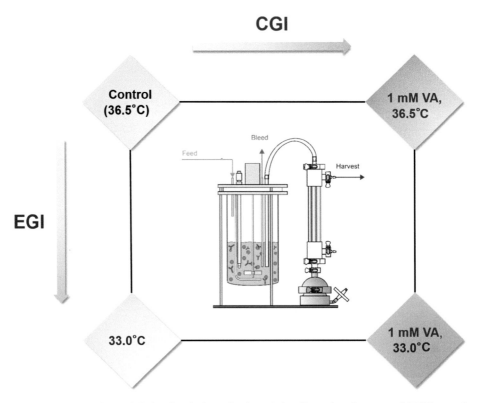

Figure 4.14 Experimental design for the investigation of the effect of environmental (EGI) growth inhibition and chemical growth inhibition (CGI).

growth inhibitor (CGI). In addition, mild hypothermia was also considered as an environmental growth inhibitor (EGI) because of its potential to reduce cell growth (Chuppa et al. 1997, Coronel et al. 2016, Kaufmann et al. 1999, Park et al. 2015). Four experiments were designed in order to investigate the effect of low temperature and valeric acid during the production phase of a perfusion bioreactor.

In particular, as shown in Figure 4.14, these included, after high cell density seeding from an N-1 perfusion bioreactor, the control culture at 36.5°C without valeric acid, a culture at the same temperature but with the addition of 1 mM valeric acid to the perfusion medium (CGI), a culture at 33.0°C (CGI) and a culture with both CGI and EGI – that is, 33.0°C – and with the same amount of valeric acid. In all cases, the perfusion runs were operated at VCD of 30×10^6 cells/mL and with a harvest rate of 1 RV/day, so the overall perfusion rate depended on the extent of growth inhibition and the resulting bleed rate. In all cases two-stage perfusion cultures were operated. Stage 1 represents the N-1 perfusion bioreactor, where cells exhibit high cell specific growth rates to rapidly achieve high cell numbers, while stage 2 represents the production phase. The separation of the two stages allows the time to steady state within the production bioreactor to be shortened.

While the combination of environmental and chemical growth inhibition did not allow a stable reactor operation, stable cell cultures were obtained at control, environmental growth inhibition and chemical growth inhibition conditions, as

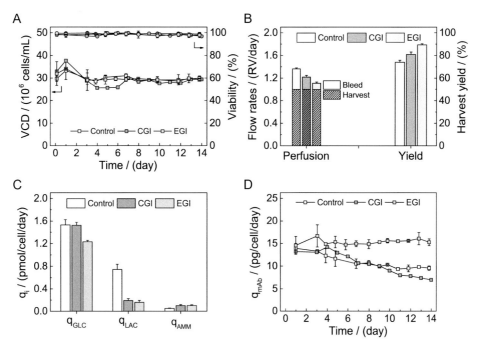

Figure 4.15 Results of the growth inhibition study. (A) VCD and viability as a function of time. (B) Flowrates and process yield for the different process conditions. (C) Cell specific rates for the different process conditions (D) Cell-specific productivity as a function of time. © Reprinted from *Biotechnology Journal*, Moritz K. F. Wolf, Aurélie Closet, Monika Bzowska, Jean-Marc Bielser, Jonathan Souquet, Hervé Broly and Massimo Morbidelli, 'Improved performance in mammalian cell perfusion cultures by growth inhibition', 2019a, with permission from John Wiley and Sons

illustrated in Figure 4.15A. The application of growth inhibition resulted in both cases in reduced bleed rates – in particular, in the case of environmental growth inhibition (Figure 4.15B). This led to a bleed rate of 0.1 RV/day, compared to 0.4 RV/day at control conditions, resulting in a process yield improvement from 74 per cent to almost 90 per cent. The different strategies affected the cellular activity. While cells exhibited similar glucose consumption under chemical growth inhibition and control conditions, under environmental inhibition conditions, cells showed a lower glucose consumption rate (Figure 4.15C). Nevertheless, under environmental and chemical growth inhibition conditions, cells showed a similar and much lower lactate production rate and also a similar but slightly higher ammonia production rate compared to the control conditions (Figure 4.15C). Considering the specific antibody productivity, cells showed the same instability in the specific protein productivity observed previously under control conditions (Karst et al. 2017a). This same trend was observed, when cells were cultivated in the presence of valeric acid (Figure 4.15D). Instead, under environmental growth inhibition, this cell-specific instability did not occur, and cell-specific mAb productivity was slightly enhanced, but more importantly, it became stable.

By investigating the cell cycle arrest, it was found (Figure 4.16) that both inhibition strategies led to cell-cycle arrest in the G0/G1 phase compared to the control condition. While environmental growth inhibition led to a significant and constant

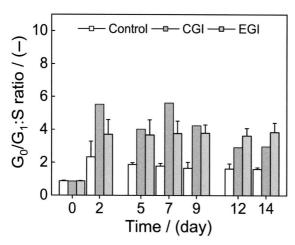

Figure 4.16 The G0/G1:S ratio represents the extent of cell-cycle arrest within the cell G0/G1 phase. Environmental and chemical growth inhibition results in a higher extent of cell-cycle arrest compared to the control conditions. While environmental growth inhibition resulted in constant cell-cycle arrest, cells seemed to adapt to the presence of the chemical growth inhibition, leading to a lower extent of cell-cycle arrest over time. © Reprinted from *Biotechnology Journal*, Moritz K. F. Wolf, Aurélie Closet, Monika Bzowska, Jean-Marc Bielser, Jonathan Souquet, Hervé Broly and Massimo Morbidelli, 'Improved performance in mammalian cell perfusion cultures by growth inhibition', 2019, with permission from John Wiley and Sons.

cell-cycle arrest, cells seemed to adapt to the presence of valeric acid when chemical growth inhibition was implemented. In particular, while an initial cell-cycle arrest was observed, after 7 days, cell proliferation started to increase again, as seen in Figure 4.16. Although the detailed mechanism of cell growth inhibition is not fully understood and it is expected to be highly cell-line dependent, these results demonstrate the potential of targeted growth inhibition during the production phase of a perfusion bioreactor.

A similar but more specific approach was chosen by Du et al. (2015) for the intensification of a non-steady-state perfusion process. Pyridopyrimidine-type molecules, which are able to arrest cells in the G0/G1 phase by target-specific binding to the cyclin-dependent kinases that controls the cell cycle, were used. By adding this molecule to the perfusion medium, the cell cycle was fully arrested (100 per cent) and cellular productivities were highly enhanced. For one cell line, the cell-specific productivity increased from 54 pg/cell/day to more than 110 pg/cell/day (Du et al. 2015). The applicability of such a molecule in a steady-state perfusion process needs to be verified, as such a process requires a minimum cellular growth in order to maintain a long-term stable culture.

Lin et al. (2017) and Wang et al. (2018) investigated different approaches to improve the cell culture media for steady-state perfusion application. In the first approach, the medium was reduced in salt components in order to further concentrate essential media ingredients and thus increase the media depth (Lin et al. 2017). In the second step, amino acid rebalancing enabled the reduction of lactate and ammonia levels and, therefore, the improvement of the cell-specific productivity without compromising viable cell density. In particular, these changes led to a cell-specific productivity increase from 40 to 80 pg/cell/day.

In a different study, Wang et al. (2018) tested the possibility of modulating and controlling cell-specific growth without adding any chemicals or supplements by investigating the impact of extracellular concentrations of sodium (Na) and potassium (K) in cell growth and productivity. Absolute potassium concentrations are involved in the regulation of the membrane potential. Changes in the membrane potential are involved in cell-cycle progression, and elevated K levels are reported to induce cell-cycle arrest (Wang et al. 2018). By reducing the Na:K ratio from 9 (control medium) to less than 1 (modified medium), the cell growth was significantly suppressed. The higher K content resulted in induced cell-cycle arrest in the G0/G1 phase. Over the time course of a culture at 40×10^6 cells/mL, Wang et al. (2018) compared the control and the modified medium. While the control medium resulted in cell-specific bleed rates between 0.3 and 0.4 RV/day, leading to significant product loss and consistent cell-specific productivities of about 20 pg/cell/day, the modified medium led to decreased bleed rates of 0.1 RV/day and consistently increased cell-specific productivities up to 115 pg/cell/day. This study further supports the conclusion that targeted cell-growth arrest improves process yields by decreasing bleed volumes, but, can also lead to elevated cell specific productivities. The approach of Wang et al. (2018), in particular, is very powerful, as Na and K are standard media components and do not require further safety tests.

4.5 PRODUCT QUALITY ATTRIBUTES: CONSIDERATIONS ON THEIR CONTROL

A fundamental issue in the production of biotherapeutics are the quality attributes of the produced molecule, which is the subject of the last step of the culture development procedure illustrated in Figure 4.5. As discussed in Chapter 1, by product quality attributes, we mean, in particular, product-related impurities and substances – such as protein aggregates and fragments, charge and glyco isoforms – that are produced from the target product as a result of events like aggregation, oxidation, deamidation, isomerisation and others (Eon-Duval et al. 2012). Other quality attributes refer to process-derived impurities such as DNA, host cell protein, raw-material-derived impurities, protein A and others. All of these have to be kept within strict specifications. We have already seen that the narrower residence-time distribution inside a perfusion bioreactor, compared to a fed-batch one, is beneficial with respect to product quality. Here, we want to discuss how to control and modulate product quality attributes in a perfusion bioreactor through the continuous addition of appropriate supplements.

In the following sub-sections, we present different scenarios where considerations on product quality play a role.

4.5.1 New and Accepted Drug Molecules

When developing a new molecule, it is important to understand how critical process parameters affect the various product quality attributes, with the values corresponding to the molecule which undergoes the clinical trials providing the reference. The relevant issue is therefore that such quality profiles are reproduced at the larger commercial scale and then maintained unchanged and consistent during the entire process life time. This is particularly challenging in the case where the process conditions have to be changed, for example, to increase productivity in the case where an existing drug is found to be active for other indications sharing the same genetic origin. In such cases, the product quality attributes have to remain within given specifications, independent of the changed process operating conditions.

4.5.2 Biosimilars

A different scenario refers to the production of a biosimilar. Here the product quality profile of the produced biosimilar has to reproduce that of the originator on the market within very narrow tolerance (Chugh & Roy 2014, Vulto & Jaquez 2017). It is important to understand how a robust manufacturing process can result in a highly similar molecule with consistent product quality. That is, glycosylation and charge variant patterns, aggregate and fragment levels and other critical attributes need to match the specifications of the originator.

4.5.3 Control of Specific Quality Attributes

As downstream purification is capable of adjusting most of the critical attributes within certain boundaries but not the N-linked glycosylation pattern, this needs to be controlled in the upstream cell culture process. N-linked glycosylation, in fact, plays a crucial role for the therapeutic efficacy and safety of the protein and, therefore, needs to be carefully kept under control.

The N-linked glycosylation pattern can be modulated both in fed-batch and perfusion cell culture processes. Villiger et al. considered a fed-batch culture and developed an optimal strategy, based on extracellular pH control, and manganese and galactose additions, to maintain a constant glycosylation pattern during the entire process (Villiger et al., 2016a, 2016b). The obtained results are illustrated in Figure 4.17, where on the left-hand side the traditional and optimised feeding strategy are compared, with the arrows indicating the main feed addition (white), the addition of manganese (light grey), the addition of galactose (dark grey) and the adjustment of the extracellular pH (black). On the right-hand side, the percentage of the different glycoforms present in the reactor in the two cases are compared as a function of time. In the optimised feeding protocol, the manganese concentration was adjusted not only immediately after the inoculation but also on days 7, 10 and 14. Additionally, the feeding of galactose on days 7, 10 and 14 was also implemented. It is remarkable that the optimal feeding strategy was developed based on a predictive mathematical model, which is further discussed in Chapter 6. Nevertheless, glycosylation patterns represent only one out of several critical product quality attributes, and if others such as charge variant patterns and fragments also need to be maintained consistently, it may become extremely difficult.

Along the same lines, and based on a similar mathematical model, Karst et al. (2017b) extended this study to perfusion processes. On the one hand, it was shown that, differently from the fed-batch case considered before, during steady-state operation in a long-term perfusion, the same glycoform distribution was produced as a function of time. On the other hand, the possibility to modulate such glycosylation profiles by properly acting on the viable cell density and the media supplement feed rates (galactose and manganese) was investigated. A preliminary set of experiments was designed to investigate the effect of such quantities on the glycoform distribution. These were executed in a benchtop bioreactor, where different steady states were achieved sequentially within a single long-term perfusion run. It was found that high ammonia levels, obtained when increasing the viable cell density, inhibited the

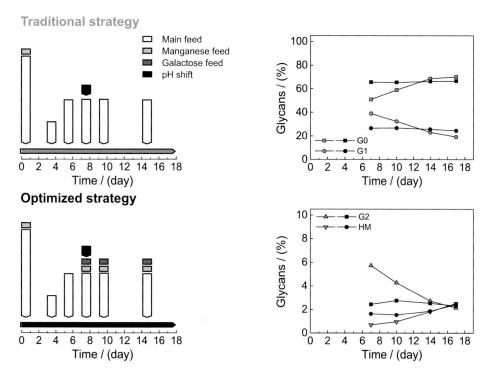

Figure 4.17 Traditional (red) and model based optimised feeding (black) strategy for a fed-batch process for N-linked glycosylation control (left). Comparison of the glycosylation pattern during the time course of the fed-batch process using the traditional (squares) and the optimised (circles) feeding strategy (right).

formation of complex galactosylated glycan structures, while the supplementation of either galactose or manganese, as well as their synergetic combination, significantly enhanced the formation of more complex glycan structures. This is illustrated by the three sequential steady states shown in Figure 4.18. These experiments were used to tune the parameters of the deterministic model, similar to the one mentioned earlier in the context of semi-batch operation, used to support the optimisation procedure. Indeed, the deterministic modelling approach combined with the sequential bioreactor runs enabled Karst et al. (2017b) to predict the system behaviour outside the reactor operating parameter space investigated experimentally, thus providing an efficient tool to develop continuous processes with controlled N-linked glycosylation patterns. This is a very interesting feature of mathematical models based on the physicochemical and biological processes occurring in a cell culture that is extensively discussed in Chapter 6.

In another case study, Bielser et al. (2019a) compared the impact of different reactor operating modes on a critical quality attribute of a conjugated protein. As shown in Figure 4.19A, this protein tends to be cut, or clipped, in two undesired pieces whose percentage represents an important quality attribute. In this work, three reactor operating modes are considered: low (LS) and high (HS) seeding fed-batch and perfusion (PF), as shown in Figure 4.19B. As expected, in both fed-batch applications, the protein accumulates in the system, and the titre increases in time. In perfusion, the protein is constantly harvested; the titre is, therefore, lower and

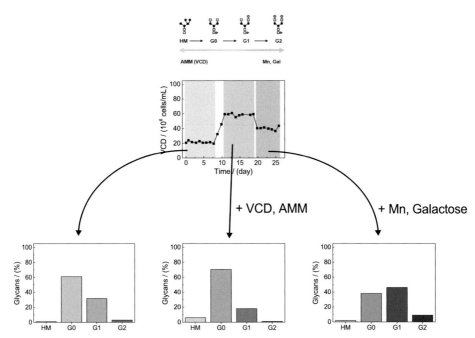

Figure 4.18 Results of the perfusion study of Karst et al. (2017b). Sequence of three steady states in a perfusion bioreactor at different VCD values: 20, 60 and 40×10^6 cells/mL (center); in the third one, a supplement of galactose and manganese was added to the feed stream. It is seen that while the increase of VCD (centre in blue) from 20 to 60×10^6 cells/mL leads to lower contents of complex glycans, the supplementation of galactose and manganese (right in red) favours more complex glycoforms. © Reprinted from *Biotechnology and Bioengineering*, Daniel J. Karst, Ernesto Scibona, Elisa Serra, Jean-Marc Bielser, Jonathan Souquet, Matthieu Stettler, Hervé Broly, Miroslav Soos, Massimo Morbidelli and Thomas K. Villiger, 'Modulation and modeling of monoclonal antibody N-linked glycosylation in mammalian cell perfusion reactors', 2017a, with permission from John Wiley and Sons

stable (Figure 4.19C). More importantly, the average residence time of the protein inside the reactor is smaller. As a result, the percentage of clipped molecules is larger in batch-type operations. In particular, in Figure 4.19D, this is seen to be 8.7 per cent and 4.9 per cent for LS and HS fed-batch, respectively, whereas it varied between 0.6 per cent and 1.5 per cent in the perfusion run. This highlights once more the potential benefit of continuous cell culture also regarding product quality, especially with the modern tendency of producing more and more complex molecules.

In summary, we have seen that perfusion bioreactors provide a very useful tool for the control of product quality and particularly the glycoform distribution, which is difficult to alter in downstream processes. Feed supplementation is very efficient in modulating glycan patterns in both fed-batch and perfusion processes. However, since perfusion allows more homogeneous or clean distributions (Karst et al. 2018, Walther et al. 2018) their modulation can be more accurate. In addition, the concept of operating the benchtop bioreactor across a sequence of steady states allows a more efficient exploration of different operating conditions – that is, one glycoform distribution per week compared to the two to three times longer times needed in the case of fed-batch operation.

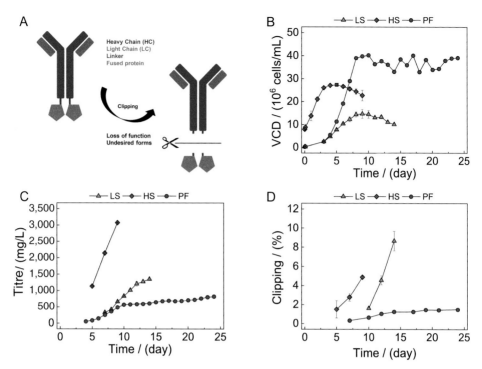

Figure 4.19 Example of a conjugated protein composed of a mAb and attached to another protein with a linker (A). Viable cell density (B), titre (C) and percentage of clipped forms (D) as a function of time for three different types of process: low (LS) and high (HS) seeding fed-batch and perfusion (PF). © Reprinted from *Journal of Biotechnology*, 302, Jean-Marc Bielser, Loïc Chappuis, Yashi Xiao, Jonathan Souquet, Hervé Broly and Massimo Morbidelli, 'Perfusion cell culture for the production of conjugated recombinant fusion proteins reduces clipping and quality heterogeneity compared to batch-mode processes', 26–31, 2019, with permission from Elsevier

4.6 PART 3: SCALE-UP TO CLINICAL AND COMMERCIAL REACTORS

Figure 4.20 schematically summarises the various steps necessary to scale to the commercial scale the perfusion culture that we have developed in the previous sections. The first one takes place at the mL scale or below, using reactor units like shake tubes, deepwell plates, and microscale bioreactors. Here, a large number of experiments is conducted with the objective of media and clone selection, but also of starting preliminary investigations of suitable process parameters for the perfusion process at the litre scale: usually the benchtop bioreactor. This is the scale where we optimise process performance, including productivity, yield and medium consumption, and make sure that all product quality parameters are within specifications. Based on our understanding of the effect of scale on processes like reactions, mass transport, mixing and others, it is, in fact, possible to conclude that this is the minimum scale for an accurate and reliable transfer of the process to the larger scales needed for clinical and commercial productions. This is largely based on the theory of multiphase reactors, which is well developed in the frame of chemical engineering and will be discussed in Chapter 5. The commercial scale required for perfusion reactors, even for expected large demands on the market such as for some monoclonal antibodies, is still relatively small, probably in the order of 500–2,000 L. These are significantly

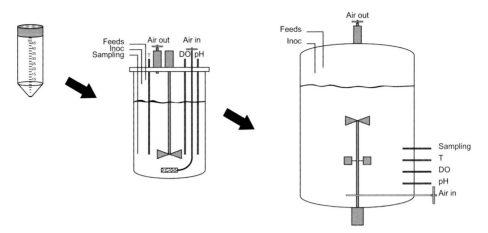

Figure 4.20 Schematic representation of the different reactor set-ups needed for the development of perfusion cultures at the commercial scale. These start from mL scale devices, such as shake tubes, move to the L scale of benchtop bioreactors and end up at the 1,000 L scale of the largest commercial units. © Reprinted from *Biochemical Engineering*, 131, Thomas K. Villiger, Benjamin Neunstoecklin, Daniel J. Karst, Eric Lucas, Matthieu Stettler, Hervé Broly, Massimo Morbidelli and Miroslav Soos, 'Experimental and CFD physical characterisation of animal cell bioreactors: From micro- to production scale', 2018, with permission from Elsevier

smaller than the counterpart fed-batch reactors that, due to their lower productivity, require larger volumes in the range of 10,000–20,000 L.

One relevant aspect, with respect to scale up of perfusion bioreactors, relates to the ancillary equipment and particularly to the cell retention device which is rather unique and specific for perfusion cultures. This involves an external loop and a filtering device, whose design and operation are scale sensitive. Mass transfer in the loop is generally not very efficient, so we need to make sure that the oxygen in the supernatant entering the external loop is not completely depleted before the loop exit. This requires sufficiently short residence times, which, however, need to be compatible with shear stress values that should not damage the cells.

4.7 CONCLUSION

In this chapter, we have presented the most important steps in order to develop and design a perfusion process for the commercial scale. This includes (1) the selection of a suitable combination of media and clones (in total, 1–2 media and 1–2 clones) with proper growth and productivity characteristics; (2) determining the $CSPR_{min}$ for the given expression systems in a suitable scale-down system and (3) validation, optimisation and fine-tuning of the process parameters at the litre scale. This represents the suitable system for further scale-up to the commercial scale. Steps 1 and 2 are preferable performed at the nL to the mL scale, while step 3 needs to be verified at the L-scale. With respect to scale-up to the commercial scale, we introduce the critical scale-up parameters and the suitable design criteria in the next chapter. The important teaching of this chapter is how to take advantage of suitable small-scale models and reduce medium consumption and work effort by following a rational approach to process development. Even if the ideal scale-down model for perfusion cultures is not yet available, the bench system is not always needed at the early stages of the design of a perfusion culture. Furthermore, ongoing technology advances will improve the current state-of-the-art scale-down models. It is only a question of when perfusion processes will be fully reproducible in the mL scale, as it is already possible for batch technologies.

5

Clinical- and Commercial-Scale Reactors

This chapter discusses the challenges in the scale-up of perfusion bioreactors from the few litres laboratory scale to the thousands of litres clinical and commercial scale. We consider comparative studies between laboratory- and large-scale reactor systems, including multiphase reactor models and computational fluidynamic tools as well as omics studies to support solid and reliable scale-up procedures. Specific scale-up issues – such as the scalability of the cell retention device long-term operation and batch definition in the case of process failure – are discussed. Finally, we evaluate the potential of single-use technologies and close the chapter with economic, financial and environmental considerations in the context of future developments in biomanufacturing.

5.1 SCALE-UP CHALLENGES AND FUTURE PERSPECTIVES

We consider here the largest scales in perfusion bioreactors, which refer to clinical and commercial production. The objective is to implement the perfusion process at such scales while retaining the same performances obtained at the litre scale that is at the end of the process development phase discussed in Chapter 4. This means the same productivity and yield, as well as the same product quality specifications. The scale-up from the small to the larger scale is a classical chemical engineering problem, as schematically illustrated in Figure 5.1. This involves issues about gas–liquid–solid mass transport, mixing and hydrodynamic stresses, which are treated in classical books about the theory of multiphase reactors (Assirelli et al. 2005, Kawase et al. 1992, Nienow 1998, Shah 1979).

The first step is indeed to understand the fluid dynamics and demonstrate comparable behavior at different scales, which is also affected by the operating mode. While the state-of-the-art fed-batch technology uses production vessels in the range of 10 to 20×10^3 L, the perfusion technology requires, in fact, comparably smaller production volumes, which, due to the higher volumetric productivity, are in the range of 0.5 to 2×10^3 L (Bielser et al. 2018, Fisher et al. 2018, Konstantinov & Cooney 2015). On the other hand, the ancillary equipment in perfusion – in particular, the cell retention device – has to be taken into account and adds complexity to the system. For both operating modes, the litre scale bioreactor is typically considered sufficient for a safe scale-up process of the stirred vessel at commercial scale. The microenvironment in large vessels is obviously much more heterogeneous in the sense that it can change substantially in space due to local differences in mixing intensity – possibly resulting in pH, oxygen concentration and temperature differences – and consequently also in

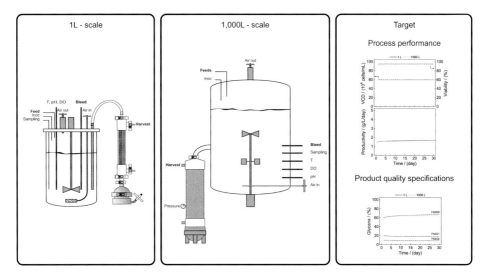

Figure 5.1 Illustration of the scale-up of a perfusion process: The process design and development end at the litre scale, and, when further increasing the scale, process conditions and equipment have to be scaled up to the commercial scale (1,000 L). Ideally, process performance (e.g., viable cell density, viability, productivity) and product quality attributes (e.g., N-linked glycosylation patterns) should remain unchanged.

different rates of oxygen uptake and carbon dioxide removal (Amanullah et al. 2001, Sieblist et al. 2011). From the point of view of a cell travelling in such a heterogeneous space, this translates into being exposed to continuously changing conditions in terms of pH, oxygen concentration and mechanical stress, which obviously affects its metabolism. In particular, in large-volume stirred vessels, the cells move in time from regions at very high shear – close to the impeller – to much calmer regions – for example, closer to the reactor ceiling – where the shear is lower. This leads to a kind of oscillatory environment, between high and low stress, which can significantly affect the cell behaviour (Neunstoecklin et al. 2015, Sieck et al. 2013).

Besides stirred vessels, which are quite common in chemical engineering operations, cell retention devices are unique and offer specific and relevant challenges. In particular, the scale-up issue results not only from the cell retention device itself but from the entire loop, where the device is inserted, which includes a pump, in the first place, but also various pipes and valves where the cells are exposed to a different and largely uncontrolled environment. In general, this may provide different types of additional stresses to the cell, related, for example, to poor temperature control, lack of oxygen or fluidynamic and mechanical conditions due to the loop geometry and the presence of an external pump. As a result, the entire design of the loop around the retention device, including the cell's residence time, becomes a critical component of the bioreactor scale-up (Voisard et al. 2003).

The issues just described become particularly challenging due to the very high cell densities in perfusion systems, which inevitably result in elevated overall production and consumption rates of metabolites and nutrients (Goudar et al. 2011). Nutrient supply and gas–liquid mass transfer may easily become limiting for the reactor operation (Mercille et al. 2000). In order to maintain the desired dissolved oxygen level in the bioreactor, the oxygen transfer rate (OTR) has to balance the cellular oxygen uptake rate (OUR). This requires a sufficiently high oxygen rate mass transfer coefficient $k_L a$ – say above 5 h^{-1} as a rule of thumb – which strongly depends on sparger type and geometry, impeller configuration, volumetric gas flowrate and agitation rate (Godoy Silva et al. 2010, Ozturk 1996). Oxygen consumption is strictly connected to the production of the undesired natural byproduct, carbon dioxide (CO_2), which is a result of the cellular metabolism and typically produced in similar molar amounts by mammalian cells (Goudar et al. 2007, Gray et al. 1996). The produced carbon dioxide affects the reactor pH, which is usually controlled through a suitable buffer system and affected by the mixing conditions. This results in a complex environment with specific levels of pH, pCO$_2$ and osmolality that need to be kept under close control since they might negatively affect cellular growth and productivity and even impact product quality, particularly when the CO_2 level increases above about 100 mmHg (Brunner et al. 2018, Gray et al. 1996, Jiang et al. 2018, Zhu et al. 2005). Higher cell densities also imply more viscous suspension cultures, which necessitate higher agitation energies to remove heterogeneities in the metabolite and gaseous species composition within the bioreactor (Ozturk 1996).

The challenges described in the preceding paragraph indicate the need for good mixing in order to make the entire cell suspension sufficiently uniform, as well as to satisfactorily deliver nutrients and exchange gaseous species. This has to comply with practical limitations coming from the stress levels that harm cell viability and a reliable transfer of the process across scales. For this, it is necessary to develop proper scale-down procedures to identify maximum operating ranges for scale-independent parameters and to mimic large-scale conditions to evaluate possible adverse effects on the other (Neunstoecklin et al. 2015). Such parameters include energy dissipation rates, mixing times and gas–liquid mass-transfer rates, which are further discussed in Section 5.2.

Before entering the details of the effect of scale on the reactor behaviour, we have to at least briefly mention some peculiarities of today's biomanufacturing, which will greatly impact the structure and design of commercial production sites in the near future (Bielser et al. 2018, Croughan et al. 2015, Karst et al. 2018, Shukla & Gottschalk 2013, Thomas 2017, Walther et al. 2016). The way we produce therapeutic proteins is at a stage of rapid evolution driven by cost pressure, access to developing markets, growing competition and clinical pipelines with new and more complex molecular formats, as well as limited manufacturing capacity, increasing regulations concerning product quality and process knowledge (such as the quality by design initiative), and the need for a higher degree of flexibility in manufacturing (Klutz et al. 2016, Siganporia et al. 2014, Walther et al. 2015, Yu et al. 2015). Process intensification through continuous and integrated operation at various stages of the production process could be the answer to some of these challenges and concerns. Continuous operation, particularly at the fermentation level, has been so far limited to the production of fragile therapeutic proteins (such as Factor VIII), and has not yet been applied in more general terms (Bödeker et al. 2013, Langer 2011). However, the possibility to integrate the different unit operations within a single end-to-end continuous production process has now been demonstrated, and its use is recommended in a number of possible production scenarios. This represents an attractive solution to downsize manufacturing units and envision more flexible production platforms (Karst et al. 2018, Klutz et al. 2015, Pollock et al. 2017, Whitford 2014). With the reduction of the equipment size, single-use applications are becoming more and more of interest. Across the industry, experts envision the manufacturing plant of the future to be fully single-use, closed and continuously operated within a modular ballroom facility, without today's large hold-up tanks (Klutz et al. 2015, Shukla & Gottschalk 2013, Whitford 2014). This obviously would not be the solution for all production situations, and many different variations will be worth considering, including hybrid technologies, with either a batch upstream process or a batch polishing unit, and only single units operated in continuous mode (Klutz et al. 2016). We come back to these aspects towards the end of this chapter to complete the picture of current and future commercial-scale manufacturing in the biopharmaceutical industry.

5.2 SCALE-UP OF STIRRED BIOREACTORS

A continuous perfusion culture typically uses a stirred tank bioreactor coupled with an internal or external cell separation device. While discussing the scalability of the cell separation device in Section 5.6, this section focuses on the scale-up of the stirred vessel. Mammalian cell cultures require specific process design and equipment layout due to their unique characteristics, mainly related to a high sensitivity to shear, to (non-physiological) carbon dioxide concentrations and to variations in the environmental conditions (Chu & Robinson 2001, Nienow 2006). In the following, we discuss critical scale-up conditions – that is, appropriate mixing, sufficient oxygen transfer and adequate stripping of carbon dioxide at low shear rate (Chisti 2001, Godoy Silva et al. 2010, Nienow 2006, Sieblist et al. 2011). We describe different approaches to combine the characterisation of the process equipment with the process knowledge in order to determine the optimal operating ranges for the perfusion culture.

Traditional scale-up strategies are rather based on empirical experience than on quantitative scientific considerations. Typically, scale-dependent parameters, such as stirring speed and gas flowrate, are adjusted at each scale during the scale-up process (Eon-Duval et al. 2012, Rouiller et al. 2012). This is a very labor-intensive approach since it requires proper ad hoc experimental activity at each scale.

On the other hand, scale-independent parameters can be defined within a reactor scale-down model and then implemented during scale up. This includes parameters such as maximum hydrodynamic stress, minimum oxygen supply and carbon dioxide removal rates, which can be used to define cell dependent ranges of operating conditions for each scale (Eon-Duval et al. 2013, Neunstoecklin et al. 2015, Sieck et al. 2014). Agitation speed and sparging intensity are valuable tools for correcting scale-up issues. However, it is necessary to address these issues systematically by performing small-scale studies that aim at simulating large-scale conditions for the investigation of possible adverse effects on cell behaviour. With respect to this, geometric similarities across different scales can provide applicable engineering rules for the determination of relevant parameters – such as the average energy dissipation rate, mixing time and gas mass-transfer rates. However, in practice, the bioreactors across the biopharmaceutical industry are often rather diverse, and this hampers the application of simple geometric scaling rules and requires either experimental measurements or computational estimation of the critical scale-up parameters (Eibl et al. 2010, Nienow et al. 2013, Villiger et al. 2018).

5.2.1 Aeration

Oxygen is a key substrate in mammalian cell cultures, particularly in the perfusion mode where the high cell densities need very large oxygen transfer rates. The low solubility of oxygen in water-based culture media makes a continuous oxygen supply necessary. The rate of oxygen supply has to balance the rate of cell oxygen consumption; otherwise, the dissolved oxygen (DO) concentration may fall below critical values – in general, in the order of 1 to 10 per cent of air saturation – resulting in changes in the cell's metabolism, which negatively impact product quality and quantity (Ozturk & Palsson 1990). However, elevated oxygen concentrations (e.g., in the case of pure oxygen sparging) may stimulate the generation of reactive oxygen species, which can alter cellular metabolism or cause cell death, as discussed in Chapters 2 and 3 (Hsie et al. 1986). For this, it is necessary to maintain cell cultures within an optimal range of DO that typically goes from 40 to 60 per cent of air saturation (Miller et al. 1987, Ozturk & Palsson 1990).

In general, animal cell lines under favorable oxygen supply conditions exhibit cell-specific oxygen consumption rates in the range of 0.05 to 0.5×10^{-12} mol/cell/h (Fleischaker & Sinskey 1981). Considering a targeted viable cell density of $1.0 - 2.0 \times 10^8$ cells/mL, the total oxygen uptake rate in a perfusion culture ranges between 5 and 100 mol/m^3/h, which is not too different from yeast or bacteria cultures. This means that mammalian cell perfusion bioreactors require high values of the oxygen transfer rate (OTR) given by the product of the volumetric mass transfer coefficient, $k_L a$ [h^{-1}], and the driving force, ΔC_{O_2}.

$$\mathrm{OTR} = k_L a \Delta C_{O_2} = k_L a (C_{O_2}^* - C_{O_2}), \qquad (5.1)$$

with $C_{O_2}^*$ representing the saturated DO concentration in the cell culture – that is, the value in equilibrium with the oxygen concentration in the gas phase – and C_{O_2} representing the actual DO concentration in the liquid phase. The maximum value of $C_{O_2}^*$ is 1 mM, corresponding to the case when pure oxygen at 1 atm is sparged into the bioreactor. The typical DO concentration in a bioreactor operation is about 50 per cent of air saturation, which corresponds to a concentration of 0.1 mM. Considering the OTR values needed in mammalian perfusion bioreactors mentioned earlier, it is seen that the mass transfer coefficient values have to be in the range of 5 up to 100 h^{-1}.

The value of $k_L a$ depends on several different factors, but mostly on the superficial velocity of the gas, V_S, and the average total energy dissipation rate $(\bar{\epsilon}_T)_g$, transferred to the liquid both through the stirrer, $(\bar{\epsilon}_T)_{lg}$, and the gas sparger $(\bar{\epsilon}_T)_s$. While the energy dissipated by the first one is defined by the mixer rotation speed and geometry, the energy delivered by the second is mainly affected by the liquid density and the superficial gas velocity, which is defined as the ratio between the volumetric gas flowrate, Q_g, and the bioreactor cross section, A:

$$V_S = \frac{Q_g}{A}. \qquad (5.2)$$

In general, with respect to the energy dissipation rate, the contribution of gas sparging is minor since low air flowrates are applied to minimise cell damage by bubble bursting. The energy input is, in fact, dominated by the stirrer under gassed conditions and is typically a function of the stirrer rotation speed and diameter, the culture density and the liquid volume inside the reactor, as discussed in Section 5.2.3.

Despite other aeration techniques – such as membrane aeration, surface aeration or the use of oxygen carriers or perfluorocarbons – bubble aeration, also referred to as sparging, is not only the most common way to introduce and remove gaseous nutrients and volatile by-products from the cell culture, but it is also the one that enables the highest transfer rates (Godoy Silva et al. 2010, Jesus & Wurm 2011, Ozturk 1996). The selection of a proper sparger device represents a compromise between maximising gas transfer rates (oxygen supply and carbon dioxide removal) while minimising cell damage and foam formation. Open pipe spargers, ring sparger and micro or frit spargers are the most commonly used aeration devices. The first one is simply a pipe ending in a nozzle through which air is introduced into the vessel, resulting in a moderate control of the bubble size, through the nozzle, diameter and position, and the gas flowrate. However, this device typically produces large bubbles, and it is therefore referred to as a macrosparger. The advantages of such devices are minimised foaming and cell death, but large bubbles have a low surface-per-unit-volume ratio and, therefore, do not support high mass-transfer rates, thus requiring high gas volumetric flowrates in order to guarantee sufficient oxygen supply. A more advanced configuration is the ring sparger, constituted by a tube with a ring shape and a number of holes of defined diameter, typically around 1 mm, allowing the simultaneous generation of multiple bubbles. On the other hand, in a microsparger, porous materials are used to create a large amount of very small bubbles in the micrometer range. Small bubbles are ideal for mass transfer because of the larger specific gas–liquid interphase surface area. However, the use of frit spargers results in reduced carbon dioxide stripping and more foam formation. Very often, a reasonable solution is to combine a frit and a large bubble sparger and use antifoaming chemicals to prevent the negative impact of foaming. In addition, shear protectants, such as Pluronic F68, can prevent cell damage caused by bubble bursting at the liquid surface (Sieblist et al. 2013). The prevention of foam formation, by antifoam addition, on the other hand, may lead to a decrease in the mass-transfer rates (Lavery & Nienow 1987).

5.2.2 Mixing

Mixing in bioreactors is needed to uniformly distribute cells, nutrients and gas bubbles – as well as the added liquids, such as base or nutritional feeds – in order to avoid spatial variations or gradients within the bioreactor, which may harm the cells. Typically, this is accomplished through the addition of energy to the cell culture broth through both mechanical agitation and gas sparging. In the ideal case, such energy should be delivered to the culture uniformly so that

1. Every single cell is exposed to a similar mixing intensity although changing in time, as seen earlier.
2. Nutrients, metabolites and by-products are uniformly distributed throughout the reactor so that the cells do not experience any variation and keep their metabolism unchanged (Godoy Silva et al. 2010, Sieblist et al. 2011).

From these considerations, it is seen that a proper design of the mixing system is needed at all scales in order to provide proper cell culture conditions while maintaining cell viability. An inadequate mixing design can, for example, allow cells to settle at the bottom of the reactor, thus creating zones with limited mass and energy transfer and a quite different environment. On the other hand, aggressive mixing is also

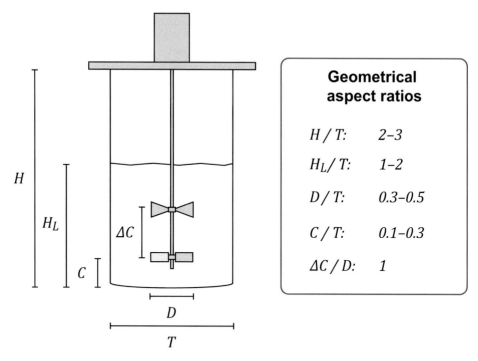

Geometrical aspect ratios

H/T:	2–3
H_L/T:	1–2
D/T:	0.3–0.5
C/T:	0.1–0.3
$\Delta C/D$:	1

Figure 5.2 Standard geometrical representation of a bioreactor set-up with the most important dimensions and aspect ratios, which have to be considered during the vessel scale-up.

negatively impacting the cells, as the resulting shear stress experienced by the cells can cause cell-growth inhibition, decrease the specific productivity or induce cell death (Marks 2003). Process and chemical engineering understanding of reactor systems with regard to mixing and power input – coupled, if needed, with the application of computational tools – are requested for smooth technology transfer across scales.

In biotech and biopharma applications, the reactor system of choice is the classical stirred vessel at all scales: from the laboratory up to the large commercial scale. Mixing within such systems is strongly affected by the vessel geometry, the presence of baffles of various shapes, and the geometry and rotation speed of the stirrer. All of this leads to different energy-dissipation rates and mixing characteristic times.

In terms of geometry (See Figure 5.2), a bioreactor can be seen as a common stirred vessel, characterised by proper dimensionless aspect ratios that have to be kept constant when aiming at geometric similarity in the reactor scale-up (Godoy Silva et al. 2010). Typical geometries include the aspect ratio of tank height (H) to tank diameter (T), equal to approximately $\frac{H}{T} = 2 - 3$. In the centre, a shaft penetrates the tank, from either the top or the bottom, connected to a motor at the outside end, while in the inside, one or more stirrers are attached to the same shaft. In the most common configurations, the aspect ratio of the impeller diameter (D) to the tank diameter is between $\frac{D}{T} = 0.3 - 0.5$. The filling height of the cell culture broth (H_L) should be around $0.7H$, resulting in an aspect ratio of filling height and tank diameter between $H_L/T = 1.0 - 2.0$. To determine the location of the stirrer, a distance from the bottom of the tank equal to $C = 0.1 - 0.3T$ appears to be a reasonable choice. In the case that more than one impeller is used in the vessel, the distance between them (ΔC) can be highly variable but is typically around one stirrer diameter.

Besides the reactor geometry, the choice of the stirrer(s) represents an important step in the reactor design, since various shapes are available leading to quite different mixing characteristics (Nienow 2010). Common designs include radial-flow stirrers such as the Rushton turbine (RT), providing a high level of intense mixing with high local shear rates, and a prevailing radial flow pattern, as well as down-pumping axial-flow stirrers such as the marine propeller, which result in an improved axial bulk movement with respect to the RT. Axial flow or pitched-blade turbines produce a flow pattern somewhere in between the others. The stirrer selection depends on several considerations, but, in particular, its specific power number (P_0), which is a function of the stirrer type itself as well as its relative size (D/T) and position in the vessel (C/T). In the case of an impeller, the power number can be used to calculate the energy dissipated into the system, P, which upon normalisation with respect to the reactor volume leads to the specific energy dissipation rate, $\bar{\epsilon}_T$.

Comparing radial-flow and axial-flow impellers, it is seen that the first ones are typically characterised by a high specific power number and consequently referred to as *high shear impellers*, while the second ones have comparably lower specific power numbers and are consequently referred to as *low/intermediate shear impellers*. Due to the shear sensitivity of animal cells, the *low/intermediate shear impellers* are often preferred for these cultures. Nevertheless, *high shear impellers*, such as the Rushton turbine, allow the breakage and complete dispersion of gas bubbles at reasonably low impeller speeds and are therefore often used at bench-scale while, e.g., (*low/intermediate shear*) marine propellers are favoured at the manufacturing scale (Nienow 2010). For all impeller types, the maximum local specific energy dissipation rate measured close to the impeller is relatively high compared to the average energy dissipation rate in the bulk, which is what we refer to as mixing heterogeneity and constitutes a crucial aspect in reactor scale-up.

In large-scale units, multiple impellers are often attached to a single-stirrer shaft so as to optimise the overall mixing performance. A typical recommended distance between two of such impellers is between one to two times the impeller diameter so that the flow patterns of the two impellers do not interfere with each other (Godoy Silva et al. 2010). Mixing effectiveness can be further improved by the implementation of baffles, which are placed perpendicular to the tank wall at 90° increments around the vessel. In the presence of baffles, large swirls can be broken, thus achieving better mixing conditions.

5.2.3 Estimation of Mixing and Mass Transfer Characteristic Times

In order to evaluate the effectiveness or characteristic time of mixing and gas–liquid mass transfer in a bioreactor, an a priori physical characterisation is necessary, which is independent of the scale and the operating mode, batch or continuous, of the bioreactor. First, we focus on the mixing effectiveness of a given set-up with a given impeller and sparging system (Karst et al. 2016, Nienow 1997). The power input (P) can be estimated experimentally as the product between the measured torque on the impeller shaft (M) and the applied agitation speed (N):

$$P = 2\pi M N, \tag{5.3}$$

from which the average energy dissipation rate is readily obtained as follows:

$$\bar{\epsilon}_T = \frac{P}{V_R}. \tag{5.4}$$

Alternatively, one can estimate the power input from the definition of the dimensionless power number P_0 under unaerated conditions, defined as

$$P_0 = \frac{P}{\rho N^3 D^5} \tag{5.5}$$

as follows:

$$P = P_0 \rho N^3 D^5, \tag{5.6}$$

where P_0 can be computed from semi-empirical correlations (Bates et al. 1966) and provides the power number for specific stirred vessel configurations as a function of the Reynolds impeller number, Re_{imp}. For example, under turbulent conditions ($Re > 2 \times 10^4$, where $Re = \frac{\rho_L u L}{\mu_L}$ with u the fluid velocity, and L the characteristic length), the power number is independent of the Reynolds number in the case of a Rushton impeller with a power number value of about 5.0 (*high shear*), while for a classical six-blade, 45° pitch turbine a value of about 1.7 is reported (*intermediate shear*) (Bujalski et al. 1987, Nienow 2010, Rutherford et al. 1996). Therefore, we can estimate the energy dissipation rate for a given impeller configuration and its specific power number as follows:

$$\bar{\epsilon}_T = \frac{P_0 \rho N^3 D^5}{V_R} \tag{5.7}$$

In order to evaluate the effect of sparging on mixing, this analysis has to be repeated in aerated conditions in order to determine the gaseous power input, P_g, which often results in a decrease of the amount of power the impeller is able to dissipate into the liquid – in particular, for high-shear conditions. This is due to a process referred to as *flooding*, consisting in the generation of air cavities around the impeller, which are stabilised by the lower pressure values prevailing in this region. Semi-empirical correlations for the estimation of the power input under aerated conditions are reported, in the case of flat-blade turbines, by Hughmark (1980).

$$\frac{P_g}{P} = 0.10 \left(\frac{Q_g}{N V_R} \right)^{-\frac{1}{4}} \left(\frac{N^2 D^4}{g D_i V_R^{\frac{2}{3}}} \right)^{-\frac{1}{5}}, \tag{5.8}$$

which has been derived for the following ranges of operating conditions:

$$0.87 < \rho_L < 1.6 \qquad \left[\frac{\text{kg}}{\text{m}^3} \right]$$

$$0.0008 < \mu_L < 0.028 \qquad [\text{Pa} \times \text{s}]$$

$$0.025 < \sigma_L < 0.072 \qquad \left[\frac{\text{N}}{\text{m}} \right]$$

$$0.1 < T < 1 \qquad [\text{m}]$$

$$0.25 < \frac{D}{T} < 0.46 \qquad [-]$$

5.2 Scale-Up of Stirred Bioreactors

$$0.31 < \frac{P_g}{P} < 0.8 \qquad [-],$$

with the volumetric gas flowrate, Q_g, the agitation speed N, the bioreactor volume V_R, the impeller diameter D, the impeller blade width D_i and the acceleration of gravity g.

The time scale of mixing is inversely proportional to the turbulent intensity. In particular, the mixing characteristic time, θ_m, can be estimated from the two following expressions (Ma et al. 2003, Nienow 1997): the first one applies when the aspect ratio of the liquid tank height to the tank diameter $\left(\frac{H_L}{T}\right)$ is equal to 1 and the second when the liquid tank height is larger than the tank diameter, so that $H_L > T$:

$$\theta_m = 5.3 \left(\frac{1}{N}\right)\left(\frac{1}{P_0^{\frac{1}{3}}}\right)\left(\frac{T}{D}\right)^2 \tag{5.9}$$

$$\theta_m = 3.3 \left(\frac{1}{N}\right)\left(\frac{1}{P_0^{\frac{1}{3}}}\right)\left(\frac{T}{D}\right)^{2.43}. \tag{5.10}$$

In order to keep constant the mixing times during scale-up, the product of $\left(\frac{1}{N}\right)\left(\frac{1}{P_0^{\frac{1}{3}}}\right)$ should be kept constant as $\frac{T}{D}$ is kept constant by definition:

$$\left(\frac{1}{N}\right)\left(\frac{1}{P_0^{\frac{1}{3}}}\right) = \text{constant}, \tag{5.11}$$

which, using Equation (5.6), leads to

$$\left(\frac{1}{N}\right)\left(\frac{\rho N D^{\frac{5}{3}}}{P^{\frac{1}{3}}}\right) = \left(\frac{\rho D^{\frac{5}{3}}}{P^{\frac{1}{3}}}\right) = \text{constant}. \tag{5.12}$$

This equation indicates that during scale-up of geometrically similar tanks (i.e., constant ratio $\frac{T}{D}$), the mixing time can be maintained constant only if the power input per unit volume is increased proportionally to the square of the impeller diameter D. However, due to hydrodynamic concerns at the large scale (as we will discuss later), these reactors tend to be designed so that the power input per unit volume is constant or decreases with the scale.

The characteristic mixing time can be measured experimentally through the dye decolorisation method, which can be applied in transparent vessels (Sieblist et al. 2016). Initially, an amount equal to 0.2 per cent of the reactor volume of a 1 weight per cent starch solution is added to the reactor, and then a 1 M iodine–potassium iodide solution is added in order to color the liquid inside the reactor. After this, a given amount of 1 M sodium thiosulphate solution is added, which reduces the iodine leading to the decolorisation of the liquid. The time between the addition of the sodium thiosulphate solution and the complete decolorisation provides an approximate estimate of the characteristic mixing time. Due to GMP regulations, this method is not applicable in production vessels.

Villiger et al. (2018) used pH-shift experiments in order to experimentally determine the characteristic mixing time. For this, they carried out pulse additions of concentrated NaOH solutions (32 per cent, 10.8 mol/L) with a volume of 0.1 per cent of the bioreactor volume. The pH was monitored every 5 seconds with two pH probes – one located at the bottom and the other one at the top of the reactor.

The starting point of the measurement was aligned with the addition of base, and the mixing characteristic time was taken as the point when both probes reached their new and equal pH value. This experiment can be repeated several times to obtain a reliable estimate of the mixing time, and, in addition, by varying agitation and aeration conditions, their effect on the mixing characteristic time can be evaluated.

Alternatively, computational fluid dynamic (CFD) approaches can be used in order to mimic *in silico* the same experiments described and compute the relative characteristic mixing times based on simulations, as discussed later in Section 5.4.

In addition to mixing effectiveness, the evaluation of gas–liquid mass transfer is of substantial importance not only to secure sufficient oxygen supply but also to remove carbon dioxide, at the different reactor scales. This is measured by the volumetric gas–liquid mass transfer coefficient $k_L a$ defined by Equation (5.1), which depends on many of the stirred vessel elements: stirrer, sparger, baffles, liquid volume and the relative operational settings such as, for example, stirring speed and gas flowrate. We have seen in the previous section that in order to support perfusion cultures with cell densities above 10^8 cells/mL, a minimum of $k_L a$ of $5\ h^{-1}$ is necessary, but higher values are definitely recommended (Ozturk 1996). For a given set-up, the corresponding $k_L a$ can be measured using the dynamic gassing method (Van't Riet 1979). First, the filled bioreactor is stripped out of all oxygen and carbon dioxide through nitrogen aeration. Next, the desired operating conditions are applied in terms of gas flowrate, agitation speed and oxygen concentration in the gas stream. By changing such conditions – for example, through a factorial design scheme – the $k_L a$ values can be calculated by fitting the measured oxygen concentration values as a function of time, as described by Bandyopadhyay et al. (1967). The obtained values are then fitted as a function of the prevailing operating conditions through the classical Van't Riet equation (1979):

$$k_L a = K \left(\frac{P_g}{V_R} \right)^\alpha V_S^\beta, \tag{5.13}$$

where, for pure water,

$$K = 0.026 \qquad \alpha = 0.4 \qquad \beta = 0.5$$

and, for ionic solutions,

$$K = 0.002 \qquad \alpha = 0.7 \qquad \beta = 0.2.$$

This equation has been derived for the following operating conditions:

$$2 \times 10^3 < V_R < 4.4 \qquad \left[m^3 \right]$$

$$500 < \frac{P}{V_R} < 10,000 \qquad \left[\frac{W}{m^3} \right],$$

with P_g denoting the gaseous power dissipated by the stirrer, V_R the volume of the bioreactor and V_S the superficial gas velocity defined by Equation (5.2). Alternatively, the expression reported by Chandrasekharan & Calderbank (1981) can be used:

$$k_L a = \frac{E}{T^4} \left(\frac{P_g}{V_R} \right)^A Q_q^{\frac{A}{\sqrt{T}}}, \tag{5.14}$$

with

$$E = 0.0248 \qquad A = 0.551.$$

This equation has been derived for the following operating conditions:

$$90 < \frac{P}{V_R} < 1700 \qquad \left[\frac{\mathrm{W}}{\mathrm{m}^3} \right]$$

$$2.6 \times 10^{-3} < V_S < 1.8 \times 10^{-2} \qquad \left[\frac{\mathrm{m}}{\mathrm{s}} \right]$$

$$0.3 < T < 1.2 \qquad [\mathrm{m}] \,,$$

with the tank diameter T, the reactor volume V_R, the gaseous power input P_g, the volumetric gas flowrate Q_g and the fitting parameters A and E.

As for the mixing effectiveness, computational fluid dynamics can also be applied to determine the local $k_L a$ values based on stirred vessel simulations, as discussed in the next section.

In general, when considering both mixing and aeration, it is important to find a maximum scale-independent variable, such as the specific energy dissipation rate, which provides sufficiently high mixing effectiveness and gas–liquid mass-transfer rates. However, such conditions should also not be harmful to the cells with respect to their shear sensitivity. Therefore, it is necessary to further analyse and evaluate the reactor fluid dynamic conditions (Nienow 1998) with respect to the shear rates to which cells are exposed at different scales.

5.3 HYDRODYNAMIC STRESS

As previously discussed, energy needs to be dissipated into a bioreactor to achieve satisfactory conditions in terms of mixing and mass transfer. However, this inevitably creates shear stresses, and we need to make sure that these are not harmful to the cells. In the frame of process scale-up, it is convenient to consider these aspects in terms of scale-independent variables, such as mass-transfer rate and overall energy dissipation rate, rather than scale-dependent variables, such as stirring speed and sparging rate (Eon-Duval et al. 2013, Neunstoecklin et al. 2015, Sieck et al. 2014). In fact, we expect to change the second ones during scale-up in such a way to keep the first ones unchanged.

In this section, we discuss the impact of shear stresses on cell viability. In general, mammalian cells show proper resistance against hydrodynamic stress, and the maximum tolerable stress threshold is highly cell line dependent (Chisti 2001, Neunstoecklin et al. 2015, Tanzeglock et al. 2009). For this, it is convenient to have a general procedure that can be used to determine such maximum tolerable threshold value for any given cell line and for any given experimental set-up. From a regulatory perspective, this is even more important and is aligned with the Quality-by-Design approach. It is fundamental to have similar hydrodynamics in bench- and large-scale manufacturing equipment to achieve similar process performance.

It is indeed difficult to determine the maximum hydrodynamic stress in multiphase systems – in particular, in bioreactor systems. Computational fluid dynamics represent an option, but in the presence of multiple phases and complex geometries, its application is not straightforward. Villiger et al. (2015) developed a simple experimental approach that can be applied to any stirred vessel at any scale and, actually, also to units of any other type, like pumps, filters or others. It is based on the use of a suspension of poly(methyl methacrylate) (PMMA) nanoparticles synthesised by monomer-starved semi-batch emulsion polymerisation (Sajjadi & Yianneskis 2003). The initial suspension is aggregated in a vessel by stirring and adding a well-defined amount of sodium chloride solution, above the critical coagulation concentration (CCC), to destabilise the suspension and promote aggregation of the nanoparticles and formation of aggregates (Moussa et al. 2007). Such aggregates of well-defined size are then diluted with deionised water to prevent any further aggregation and to obtain a stable aggregate suspension. Now, the idea is that these aggregates are fragile and, if exposed to a certain shear stress, can break, and actually, the higher the stress they are exposed to, the smaller the size they reach. It follows that such suspensions can be introduced in any reactor or device of other sort and circulated for a sufficiently long time so that breakage can occur until steady-state dimensions are reached. Their final size can then be measured, for example, by small-angle light scattering (SALS), and this provides a measurement of the maximum shear rate that the aggregates have experienced inside the bioreactor. Clearly, this does not tell us anything about where in the unit such maximum shear occurs, but only what is the maximum shear experienced by the aggregates (or, equivalently, by the cells) somewhere in the device. To obtain a quantitative estimate of this, we obviously need a calibration curve to go from the measured aggregate size to the value of the maximum shear experienced, which Moussa et al. (2007) provided using a dedicated device. In particular, they used the axisymmetric extensional flow generated at the entrance of contracting nozzles of various diameters and flowrates (Saha et al. 2014). The corresponding shear rate was computed by computational fluid dynamics (Soos et al. 2010). Thus, by correlating the computed shear values with the aggregate sizes measured in the same device when circulating the PMMA aggregate suspension, the desired calibration curve could be constructed. This technique was used in a series of experiments performed in classical stirred tank reactors with given tank diameter, Rushton turbine impeller and a four-element baffle configuration, for various impeller rotation speeds and air flowrates. The obtained values of the maximum effective hydrodynamic stress, τ_{max}, for a given stirred tank reactor at given operating conditions were in good agreement with literature data (Neunstoecklin et al. 2015, Villiger et al. 2015).

Neunstoecklin et al. (2016) used this experimental approach to determine the relative importance of stirring and sparging to the maximum hydrodynamic stress in a 300 L bioreactor. For this, τ_{max} was measured for various combinations of stirring and sparging systems. In particular, two types of ring spargers, located below a Rushton impeller, were considered: one with eight sintered spargers (average pore diameter equal to 50 μm) and the other with eight nozzles with a diameter of 1 mm. The obtained results are shown in Figure 5.3 for different sparging flowrates as a function of the stirring speed, represented in terms of the Reynolds number, defined as $Re_{imp} = \frac{\rho_L N D^2}{\mu_L}$. It is seen that the maximum hydrodynamic stress for non-aerated conditions increases linearly with increasing Re_{imp}, in a log–log scale. When considering low gas flowrates (i.e., 0.03 vvm) the measured values of τ_{max} when using the sintered sparger are comparable to those measured for single-phase conditions, indicating the negligible contribution of small bubbles. In contrast,

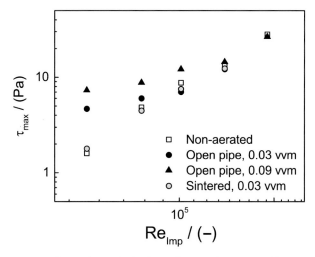

Figure 5.3 Maximum hydrodynamic stress in a 300 L bioreactor measured by the shear-sensitive PMMA system developed by Villiger et al. (2015). Two types of spargers, one with 1 mm open nozzles and the other with a 50-μm sinter sparger, were investigated at various agitation speeds and gas flowrates. Open squares correspond to the non-aerated single-phase condition. The applied gas flowrates for the nozzle sparger were equal to 0.03 vvm (filled circle) and 0.09 vvm (filled triangle), while for the 50-μm sparger, the gas flowrate was equal to 0.03 vvm (grey circle). © Reprinted by permission from Springer Nature, *Applied Microbiology and Biotechnology*, 'Pilot-scale verification of maximum tolerable hydrodynamic stress for mammalian cell culture', Benjamin Neunstoecklin, Thomas K. Villiger, Eric Lucas, Matthieu Stettler, Hervé Broly, Massimo Morbidelli and Miroslav Soos, 2016

when using the 1 mm open pipe system, which generates larger bubbles, the value of τ_{max} increases at low Re_{imp} to about 5 Pa with gas flowrates of 0.03 vvm and even more, at about 8 Pa, for larger gas flowrates of 0.09 vvm. By increasing the impeller speed, the value of τ_{max} increases in all cases, but the effect of sparging decreases and becomes negligible at sufficiently large values of Re_{imp} where the maximum hydrodynamic stress is dominated by the turbulence generated by the impeller (Neunstoecklin et al. 2016).

Neunstoecklin et al. (2015) used this technique to determine the maximum operating hydrodynamic stress for mammalian cell cultures for different cell lines (CHO and Sp2/0). The authors realised a scale-down model to determine the threshold values of hydrodynamic stress, which was evaluated based on different cell characteristics – that is, cell viability, growth, morphology, metabolism and productivity. In particular, a modified bioreactor set-up was used in order to expose the cells to different and well-defined hydrodynamic stresses, as illustrated in Figure 5.4. In addition to the typical stirred vessel, this set-up is equipped with an external loop, driven by a contact-free centrifugal pump, containing a nozzle. Pumping the cell culture liquid through the external loop with the nozzle, an oscillating hydrodynamic stress is imposed to the cells, similar to what they experience in a large scale bioreactor, with the high stresses in the nozzle representing the regions in the vicinity of the stirrer and the lower stresses in the stirred vessel representing the bulk zone away from the stirred region (Soos et al. 2013). The exposure time in the loop was set to 90 seconds, which was correlated to the cells' exposure time in zones with large hydrodynamic stresses

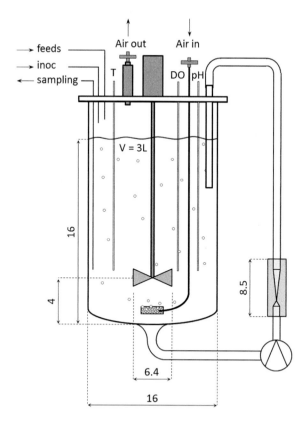

Figure 5.4 Set-up and geometry of the 3 L bioreactor system used by Neunstoecklin et al. (2015) to generate an oscillating hydrodynamic stress on the cells. Nozzles of various diameters were used for the generation of different shear rate maxima. All given values are in cm.
© Reprinted by permission from Springer Nature, *Applied Microbiology and Biotechnology*, Neunstoecklin, B., Stettler, M., Solacroup, T., Broly, H., Morbidelli, M. & Soos, M. (2015), 'Determination of the maximum operating range of hydrodynamic stress in mammalian cell culture', Journal of Biotechnology 194, 100–109.

in an in-house 5,000 L bioreactor. By varying the nozzle diameter, hydrodynamic stresses of different magnitudes can be imposed. Thus, by circulating the PMMA aggregate suspension inside the unit and measuring the corresponding equilibrium size, using the calibration curve, as mentioned before, the maximum shear experienced by the aggregates (and also by the cells, since they exhibit very similar size) inside the unit could be estimated.

The decrease in cell viability measured after 24 h of culture in bioreactors sparged with nozzles of different sizes are shown as a function of the maximum measured shear stress in Figure 5.5 for the two considered cell lines. It is seen that cell viability is not affected by shear stress values below a critical value, above which cell viability decreases.

This behaviour is common for both cell lines, with the maximum tolerable threshold value at τ_{max} of 32 ± 4 Pa for the CHO cell line and at 25 ± 2 Pa for the Sp2/0 cell line. Neunstoecklin et al. (2016) performed validation experiments for the given Sp2/0 cell line in a much larger 300 L pilot-scale bioreactor operated at different agitation and sparging rates to mimic hydrodynamic stress conditions in the range from 7 to 28 Pa and confirmed that, independent of the reactor scale, growth behaviour, metabolite concentrations, productivity and product quality showed a dependency on the different environmental stress conditions. Although this study was performed in fed-batch cultivations, similar concepts are also expected to apply in perfusion conditions. The maximum tolerable stress, τ_{max}, is, in fact, a cell-specific

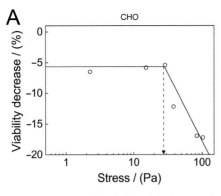
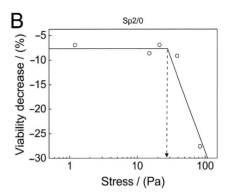

Figure 5.5 Example of the critical stress determination (dashed arrows) based on the viability drop during the first 24 h of the culture for a CHO (A) and a Sp2/0 cell line (B). © Reprinted from *Journal of Biotechnology*, 194, Benjamin Neunstoecklin, Matthieu Stettler, Thomas Solacroup, Hervé Broly, Massimo Morbidelli and Miroslav Soos, 'Determination of the maximum operating range of hydrodynamic stress in mammalian cell culture', 100–109, 2015, with permission from Elsevier

quantity and not a process-specific one, and therefore, it is expected to hold true for both batch and continuous operations and across different scales.

5.4 COMPUTATIONAL FLUID DYNAMICS

We have seen the importance of characterising bioreactors at different scales in terms of mixing, aeration and shear stress and discussed several experimental techniques to measure the corresponding relevant quantities. In this section, we briefly discuss the possibility of determining the same quantities through numerical simulations – that is, using computational fluid dynamics (CFD). The main advantage of CFD is to provide estimates of all local quantities in the entire reactor volume, thus providing a full description of the reactor heterogeneity, in time and space, which is particularly relevant for commercial scale bioreactors (Villiger et al. 2018). Computational fluid dynamics provides a valuable tool to evaluate this heterogeneity for various reactor systems of various geometries and sizes (Ranganathan & Sivaraman 2011) and at different operating conditions.

This methodology applies numerical techniques to solve the equations describing the motion of fluids within the bioreactor. The Navier–Stokes equations build the basis of CFD containing expressions for the conservation of mass, momentum and energy. The use of discretisation techniques enables the solving of the different equations for a finite set of volumes, so-called cells, and results in discrete flow fields. Assumptions, such as steady-state or inviscid flow, simplify the solutions and accelerate the convergence of the solutions (Sharma et al. 2011).

A typical CFD analysis includes three stages.

1. Building the model using appropriate meshing and boundary conditions, as well as information about the fluid properties.
2. Solving the various momentum, mass and energy balance equations by discretisation and producing results in some convenient form to be analysed.
3. Interpreting the data in the form of vectors and contour plots to extract the desired information (Versteeg & Malalasekera 1995).

In particular, the comparison of calculated CFD and experimentally determined values can be used for the validation of the obtained results. There is various commercial CFD software available for computational fluid dynamic simulations, which greatly simplify the last two steps (Sharma et al. 2011). This is relatively advanced software that allows one to select proper models for the bioreactor, simulate various geometries and the rotation of different impellers, and model the laminar or turbulent conditions inside the reactor. The software calculates the resulting flow field, with a bioreactor-size-dependent mesh size. Furthermore, it is possible to model pressure, momentum, turbulent kinetic energy and dissipation rates, thus computing the time and space distribution of hydrodynamic stresses. CFD can also be used to simulate the trajectory of a massless tracer, thus, for example, reproducing some of the mixing experiment described earlier and then comparing computed and experimental characteristic mixing times. Among the most common applications of CFD, we can list the determination of $k_L a$ values within different type of bioreactors, hydrodynamic forces within different bioreactor regimes, local energy dissipation rates and optimum impeller locations by estimating the different hydrodynamic properties and different flow regimes (Ahmed et al. 2010, Devi & Kumar 2017, Kelly 2008, Ranganathan & Sivaraman 2011, Soos et al. 2013). Of course, depending on the level of detail of the desired simulation and the complexity of the system in terms of number of phases, steady-state or transient behaviour, and overall size of the object to be simulated, the required computation time may grow significantly and become unfeasible. For this, various techniques to reduce the size of the process can be introduced – for example, using pseudo-phases or partitioning the entire reactor volume in separate zones. In principle, CFD can be successfully applied not only to gain fundamental understanding of different processes taking place in a given bioreactor configuration but also to optimise the operating conditions and bioreactor design during scale-up.

5.5 CHARACTERISATION OF THE BIOREACTOR: CASE STUDIES

In this section, we illustrate some case studies where the concepts discussed earlier have been applied to perform successful process scale-up from the bench to the manufacturing scale. In the first part, we discuss a comparative characterisation study of different reactor scales from the mL to the commercial scale in terms of mixing, aeration and hydrodynamic stress using both experimental and computational tools (Villiger et al. 2018). Next, using the results of an omics study by Bertrand et al. (2018), we compare cultures at different scales not only in terms of productivity and product quality but also at the level of intracellular processes. Finally, we describe the full characterisation of a perfusion set-up before proceeding with the discussion about the scale-up of cell retention devices in the next section.

Villiger et al. (2018) performed an experimental and computational comparative study of different animal cell culture bioreactors ranging from the micro- to the production scale. The study focused on the comparison of the fermenters and did not consider any auxiliary equipment, such as the cell retention device. The considered reactors are schematically illustrated in Figure 5.6 and include the disposable 15 mL ambr micro-bioreactor, two different 3 L scale glass vessel reactors (with different impellers) and the larger stainless-steel bioreactors with volume of 270 L, 5,000 L and 15,000 L. The comparison included the experimental determination of the

5.5 Characterisation of the Bioreactor: Case Studies

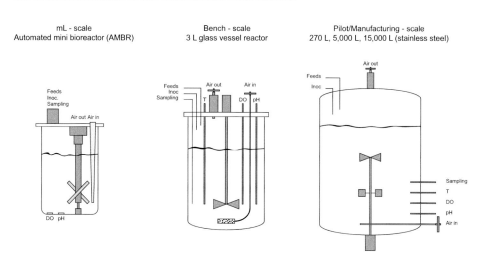

mL - scale	Bench - scale	Pilot/Manufacturing - scale
Automated mini bioreactor (AMBR)	3 L glass vessel reactor	270 L, 5,000 L, 15,000 L (stainless steel)

Figure 5.6 Schematic representation of the different reactor set-ups ranging from the mL-scale (ambr© 15 system) with nominal working volume of 10 mL, to the bench-scale with a nominal working volume of 3 L to the manufacturing scale with nominal working volumes of 270, 5,000, and 15,000 L. © Reprinted from *Biochemical Engineering Journal*, 131, Thomas K. Villiger, Benjamin Neunstoecklin, Daniel J. Karst, Eric Lucas, Matthieu Stettler, Hervé Broly, Massimo Morbidelli and Miroslav Soos, 'Experimental and CFD physical characterisation of animal cell bioreactors: From micro- to production scale', 2018, with permission from Elsevier

maximum hydrodynamic stresses, the mixing time, the volumetric gas–liquid mass transfer coefficient and the dissolved oxygen concentration at different locations in the reactor volume. The same quantities were also computed through computational fluid dynamic simulations in order to evaluate their heterogeneity in the reactor. Both experimental and computational results revealed that although the volumetric power input number of 13.5 W/m^3 is kept constant, there are quite different shear stress distributions and maximum values across scales. The difference in the hydrodynamic stress patterns was mainly due to different reactor geometries. Highest stress values are located near the impeller, and lower values are located above the impeller. In particular, for the ambr© 15 system, the hydrodynamic stress distributions calculated by CFD revealed that stress values are rather low in a significant fraction of the reactor volume due to the position of the stirrer very close to the bottom, which resulted in poorly mixed regions at the top of the reactor. The maximum stress values were measured or estimated (for 5,000 and 15,000 L) at each scale. It was found that the maximum values of the energy dissipation rates systemically increase with the volume, although they actually appear only in a very small region within the reactor volume. With respect to the mixing time, the obtained values increase with the reactor volume from several seconds in the 3 L system to almost two minutes in the 15,000 L bioreactor for the same specific energy dissipation rate. As observed by Villiger et al. (2018), longer mixing times are an inevitable consequence of increasing the scale. On the other hand, it is key to keep heterogeneity to a minimum in order to avoid the cells' exposure to non-physiological conditions during acidic or basic feed additions. This is even truer for continuous cultivations, as the continuous media exchange requires a full homogenisation of the cell culture broth. Accordingly, if critical and tolerated by the cells, the power input should be increased at the manufacturing scale in order to improve the reactor homogeneity.

Oxygen transfer is of crucial importance for a healthy culture. This has been investigated by combining a CFD simulation of the cell suspension with a population balance equation describing the behaviour of gas bubbles of different sizes and eventually computing the average value of the $k_L a$. Also, in this case, quite heterogeneous patterns and values have been found between different scales. The results highlight the expected strong impact of the bioreactor configuration in terms of sparger, baffles, probes and agitation on the $k_L a$. The study of Villiger et al. (2018) shows that modelling can be very efficient to identify operating conditions, primarily the energy dissipation rate, that optimise the reactor behaviour, for example, by mitigating inhomogeneity and gas mass transfer issues in the large scale. However, in order to be effective, this approach requires a clear knowledge of the shear stress threshold of the cell line, as discussed earlier (Neunstoecklin et al. 2015 and 2016).

Bertrand et al. (2018) performed a proteomic analysis in both 15 mL microbioreactors and a 300 L pilot-scale bioreactor for an industrial fed-batch platform to investigate the impact of the reactor scale on process performance, with particular reference to the intracellular processes. For this, a 17-day fed-batch CHO culture was considered. Consistency between the different scales was ensured by using similar inoculum cell densities of 0.2×10^6 cells/mL, equal standard cultivation conditions (36.5°C, pH 7.15, DO of 40 per cent air saturation) and similar feeding strategies. The time evolution of the proteome was measured, and nearly absolute accordance between the scales was verified by data-mining methods, such as hierarchical clustering, and in-detail analysis on a single protein base. While similar cell density profiles, growth rates and specific production and consumption rates were measured as a function of time independent of the reactor scale, small differences were observed with respect to the principal enzymes related to N-glycosylation resulting in slightly different glycan patterns. This could be a result of the differences in geometry, shear rate, mixing and oxygen mass transfer patterns between the ambr© 15 and the 300 L pilot scale. The significant fraction of poorly mixed regions within the ambr system, observed by Villiger et al. (2018), may play a significant role here. Nevertheless, it should be mentioned that the observed differences had a minor impact on the glycan distribution, and a highly comparable process performance overall was observed in the two reactors, thus confirming the reliability of the scale-up procedures.

A rather complete characterisation of a perfusion bioreactor – in terms of hydrodynamic shear rate, $k_L a$, and mixing time – was reported by Karst et al. (2016) in the frame of a study intended to compare ATF and TFF retention devices. The ATF pump was periodically displacing 100 mL at a given flowrate of 1–1.5 L/min. The TFF set-up was driven by a bearingless centrifugal pump attached to the hollow fibre, as already discussed in Chapter 2. The reactor system was a 2.5 L benchtop bioreactor with 13 cm diameter (T). In the case of the TFF system, the glass vessel was modified with a bottom outlet to enable the direct attachment of the bearingless pump to circulate the cell culture through the external loop. A six-blade Rushton turbine impeller with a diameter, D, of 4.5 cm was used, and a four-whole-open-pipe sparger with 1 mm openings was installed 1 cm below the impeller to supply cells with the required amount of oxygen. The working volume was set to 1.5 L. The authors evaluated mass transfer coefficients for the given set-up as a function of the specific power input and superficial gas velocity at various stirring speeds (300–500 rpm), and sparging rates (0.11 – 0.33 vvm). For the given bioreactor configurations, $k_L a$ values in the range of $5 - 40$ h^{-1} were determined, which would be sufficient to support up to 10^8 cells/mL. The

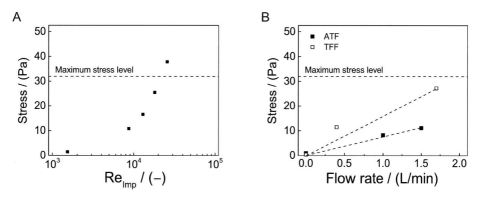

Figure 5.7 Maximum hydrodynamic stress measured in a 2 L bioreactor: (A) as a function of the impeller Reynolds number (Re_{imp}) and with no cell filtration and no sparging; (B) as a function of the recirculation flowrate in the filter in the presence of a ATF (filled squares) and a TFF (open squares) cell retention device with a sparging rate of 0.22 vvm and an agitation speed of 400 rpm. The horizontal dashed line indicates the critical shear stress value of 32 Pa for the considered CHO cell line. © Reprinted from *Biochemical Engineering Journal*, 110, Daniel J. Karst, Elisa Serra, Thomas K. Villiger, Miroslav Soos and Massimo Morbidelli, 'Characterization and comparison of ATF and TFF in stirred bioreactors for continuous mammalian cell culture processes', 2016, with permission from Elsevier

mixing times were estimated from the power number computed based on the power absorbed by the motor operating the stirrer shaft, using the empirical correlation (Equation (5.9)). The obtained values for all conditions considered were in the range of 2–6 seconds, which is sufficiently fast to avoid inhomogeneities in the system. Finally, the effective maximum hydrodynamic stress within the system was estimated using the shear sensitive PMMA particulate system illustrated above (Villiger et al. 2015). The maximum tolerable stress for the used cell line was known to be about 32 Pa. Karst et al. (2016a) evaluated the individual contribution of the filtration module on the maximum hydrodynamic stress. In the case of no filtration module and no sparging, a maximum shear stress of 37 Pa – that is larger than the one tolerated by the cells – was measured at 500 rpm, as shown in Figure 5.7A. In addition, taking the cell retention system into account, the maximum shear stress value was measured at a sparging rate of 0.22 vvm, with an agitation speed of 400 rpm as a function of the circulation flowrate in the ATF or TFF retention device, as shown in Figure 5.7B. It was found that the hydrodynamic stress to which cells were exposed during cultivation was smaller in the ATF compared to the TFF system at comparable flowrates, and the resulting maximum hydrodynamic stresses were below the reported threshold for the given cell line. Therefore, a safe culture could be expected for the entire operating range. By comparing the cell culture performance within the two set-ups, very similar culture performance and product quality were observed. However, with respect to product retention, the use of the TFF resulted in considerable protein retention compared to ATF, as discussed in Chapter 2.

From the results of this study, it clearly appears that a proper reactor characterisation is of fundamental importance for process design and scale-up, particularly with reference to the shear stress. The cell retention device brings an additional contribution to the shear stress present in the reactor system. Wang et al. (2017) found for a TFF system driven by a peristaltic pump significantly higher shear rates

Cell specific maximum tolerable stress, τ_{max}

CFD and experimental reactor characterization

Power drawn into the system

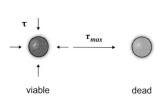

viable   dead

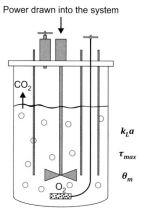

CO_2

$k_L a$

τ_{max}

θ_m

O_2

Figure 5.8 Example of the critical stress determination (dashed arrows) based on the viability drop during the first 24 h of the culture for a CHO (A) and a Sp2/0 cell line (B). Key characterisation for the reliable scale-up of a perfusion process: (1) determination of the cell specific maximum tolerable dynamic stress (Neunstoecklin et al. 2015) and (2) computational and experimental characterisation of the reactor system in terms of volumetric gas–liquid mass-transfer rate, $k_L a$, mixing characteristic time, θ_m, and maximum shear rate, τ_{max}.

compared to the ATF system or to the TFF system driven by a bearingless centrifugal pump, resulting in a higher degree of cell death and product retention. With respect to scale-up, characterisation studies for commercial-scale perfusion bioreactors are currently not available, but with the methods described in this chapter, it is definitely possible to ensure reliable scale-up to commercial scale.

The critical steps for a reliable scale-up are summarised pictorially in Figure 5.8. These include the preliminary determination of the maximum tolerable hydrodynamic stress for the given cell line and then the physical characterisation of the bioreactors at different scales in terms of interphase mass transfer, mixing and maximum shear rate.

5.6   CELL RETENTION AT LARGE SCALE

Robust and scalable cell retention devices have been a major technological difficulty that limited the large-scale application of perfusion processes for a long time. Today, commercial solutions are available at lab and large scale mainly based on tangential filtration. This is the preferred separation technology because it offers full cell retention. In addition, one of the main technical challenges is that complex mixtures such as cell cultures favour membrane fouling, and tangential flow mitigates this undesired effect. Recent technological innovations lie mostly with the filter design, either in its shape or in its operating mode. For example, filters can be designed as flat sheets or hollow fibres, to maximise the contact surface area per unit volume. The recirculation flow can be either constant and unidirectional as in TFF or it can be alternating – that is, regularly inverting its direction at a constant frequency as in ATF (Figure 5.9). Next, we address some scale-up considerations for these two filtration technologies.

5.6 Cell Retention at Large Scale

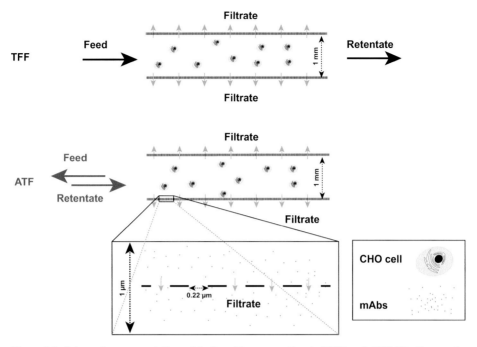

Figure 5.9 Schematic representation of hollow fibres operating in TFF and ATF filtration mode. The filter is the same, but the flow direction is either unidirectional and constant (TFF) or inverts direction at a regular frequency (ATF). The filtration direction shown by the blue arrows is perpendicular to the flow inside the fibres. The zoom on the membrane surface compares the size of the mAbs (assumed to be 10 nm large) to the pores of the membrane. The mammalian cell diameters typically range between 10 and 20 μm and are not represented at scale.

In order to scale a perfusion culture from the laboratory to the manufacturing scale, there are a few key parameters that must be carefully controlled. For example, the membrane material and porosity (usually 0.22 μm in polysulfone or polyether-sulfone) must be the same. The two main parameters to be controlled – that is, kept substantially constant – in scaling up are the residence time of the cell culture in the filtration device and the specific filtration capacity in volume of permeate per surface area of membrane and per unit time. The first one is due to the fact that the environment in the filtration device is not controlled, and therefore, the oxygen and nutrient concentrations, pH and even temperature may change with no possibility for any control action to intervene or to correct such change. The danger is that some conditions are created that affect cell viability, like complete depletion of dissolved oxygen, and this requires the cell residence time to be controlled in the filtration device. The second parameter is related to the limited filtration capacity of the membrane, which is particularly challenged at high cell densities like in perfusion cultures. The additional shear stress experienced by the cells in the filtration loop should also be taken into consideration, as discussed in the context of Figure 5.7.

Commercial ATF systems are available for different bioreactor scales ranging from a few litres up to 1,000 L at a perfusion rate in the order of 1 RV/day. The perfusion volumes can be increased by adding filtration units and running them in parallel. The design of the filtration units for each scale, including the pump head, is

Table 5.1 Comparison of a 2 L bioreactor with perfusion rate of 1 RV/day and equipped with an ATF2 unit (filtration surface area 0.13 m^2) and a 1,000 L bioreactor with ATF10 (11 m^2) and various perfusion rates. The third column reports the flowrates of the recirculation loop shown in Figure 5.10 needed to have the same filtration capacity.

Filtration capacity / L/m^2/day	ATF2 permeate flowrate for a 2 L bioreactor / L/day	Recirculation loop flowrate in Figure 5.10 / L/day	ATF10 permeate flowrate for a 1,000 L bioreactor / L/day
15.4	2	0	170
45.4	5.9	3.9	500
90.8	11.8	9.8	1,000

Figure 5.10 Perfusion bioreactor set-up with a recirculation loop to artificially increase the permeate flux through the membrane without affecting the perfusion rate. An example of how this flux can be used is described in Table 5.1.
© Reprinted from *Biochemical Engineering Journal*, 110, Daniel J. Karst, Elisa Serra, Thomas K. Villiger, Miroslav Soos and Massimo Morbidelli, 'Characterization and comparison of ATF and TFF in stirred bioreactors for continuous mammalian cell culture processes', 2016, with permission from Elsevier

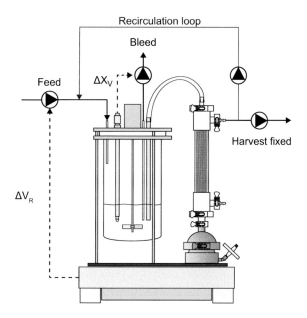

standardised and guidelines for the scale-up are usually suggested by the manufacturers. To implement ATF at commercial scale, the following parameters should be controlled:

1. Filtration capacity

   The filtration capacity is given by the ratio of the permeate (harvest) volumetric flowrate and the membrane total surface area per unit of time. Because of the standardised design of the ATF, this is usually significantly oversized at the lab scale, since the total permeate flow per surface membrane is typically very low. One option to test the filtration capacity at low scale but in conditions representative of the manufacturing ones is to use a recirculation loop to pump the excess of harvest volume back into the bioreactor without affecting the perfusion rate, as shown in Figure 5.10. The application of this concept is illustrated in Table 5.1 with reference to a 2 L bioreactor equipped with an ATF2 device exhibiting a filtration surface area equal to 0.13 m^2 and operated at a perfusion rate of 1 RV/day. If we want to reproduce the same filtration capacity values listed in the first column of

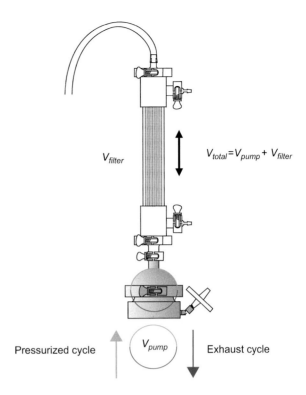

Figure 5.11 Schematic of a lab-scale ATF device showing the position of the membrane during the pressurisation cycle, when the liquid is pushed back into the bioreactor, and the exhaust cycle, when the liquid is sucked into the filtration unit. For the residence time calculation, the ratio of the volume of the pump and the total volume is considered, as described in Equation (5.15).

the table, as in a 1,000 L bioreactor equipped with an ATF10 device exhibiting a filtration surface area of 11 m², operated at different perfusion rates (last column in the table), we need to apply the permeate and external loop flowrates listed in the second and third columns of the table, respectively. When testing this parameter before scale-up, the same duration and total amount of permeate per surface area as in the large scale should be considered.

2. Residence time

The average residence time in a vessel of any geometry operated at steady state can be calculated as the ratio between the vessel volume and the inlet (or outlet) volumetric flowrate. For the ATF, this depends on the alternating cycle frequency and on the volume exchange ratio, defined here as the ratio between the displaced volume inside the device, equal to the pump head volume (Figure 5.11), and the total ATF volume. This ratio is typically close to one because the pump head volume is significantly higher than the volume of the hollow fibres and the connection tubing (that are never empty) and therefore accounts for most of the total volume. Accordingly, for an ATF unit, the average residence time, which is intended to be kept constant during scale-up across scales, is given by

$$\tau = \frac{V}{Q} = \frac{V_{total}}{V_{pump} \times f}. \tag{5.15}$$

3. Shear stress

Shear stress mainly depends on the alternating flow velocity inside the fibres, which should be kept constant across scales. This comes, of course, in addition to the pump specific shear rate.

4. Backflush

During the operation of an ATF unit, it is possible that the permeate flux changes direction; this is referred to as backflush, where some permeate goes back inside the hollow fibres. This has a beneficial effect on the filter operation since it helps to maintain the cleanliness of the membrane by reducing fouling. The permeate flux direction and flowrate depend on the transmembrane pressure (TMP) – that is, the pressure difference between the retentate and the permeate side of the hollow–fibre membrane. In normal operation, the pressure gradient decreases towards the permeate side, but when this is reversed, then the flux direction changes, leading to the backflush. In ATF, because of the alternating operating mode, the retentate flux continuously changes direction and velocity during each cycle, and therefore, the TMP changes accordingly along the hollow fibre and in time. With a higher circulation rate in the ATF, the maximum tangential velocity is larger, and the TMP gradients are also larger, resulting in a stronger backflush. Because of the design of the ATF, the backflush is constant if the ratio between the permeate rate and the ATF rate remains constant. Therefore, the external recirculation loop shown in Figure 5.10 can also be used to change the ratio between the ATF rate and the permeate flowrate to reproduce the manufacturing conditions.

5. Filtration device–to–reactor volume ratio

This ratio is proportional to the average number of times that a cell travels through the retention device, and it should not change significantly across scales. This is probably the most difficult parameter to mimic at small scale, where it tends to be very large. Since this is the worst case with respect to cell behaviour, the manufacturing scale is usually not impacted by this parameter. In general, the most advisable strategy for the reliable scale-up of an ATF device is to keep constant the length of the hollow fibres while properly increasing their number. This enables constant pressure drop across the filtration package, which in turn implies constant liquid flowrate, constant average residence time and constant shear stress. On the other hand, increasing the fibre length (with constant number) would lead to higher back pressures and consequently larger shear stresses. This effect was demonstrated at lab-scale by Karst et al. (2016).

Commercial TFF systems, equipped with some devices for the control of the recirculation and the harvest streams, are available on the market for lab-scale bioreactors. At the manufacturing scale, filters like the Prostak^{TM} flat sheet from Millipore offer the possibility to have enough filtration surface, but the recirculation system is not provided. An external skid, including all the necessary pumps and control systems to monitor and control the recirculation loop, must be designed ad hoc. To implement TFF at manufacturing scale, the following parameters should be controlled:

1. Filtration capacity

This parameter should be tested at the small scale, before scale-up, for the same duration considered for the bioreactor design and for the same amount of permeate per surface area.

2. Residence time

The average residence time can easily be computed as the ratio between the total free volume of the filtration loop and the circulating volumetric flowrate (Figure 5.13).

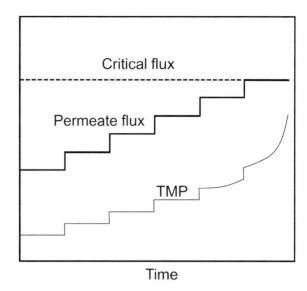

Figure 5.12 Experimental procedure to define the critical flux. The permeate flux is increased stepwise, and the corresponding TMP increase is monitored. The critical flux is reached when the TMP is not stable anymore and increases strongly due to irreversible fouling.

3. Shear stress

Shear stress should be minimised by avoiding all possible sources of pressure drop in the circulation loop, like valves, changes in cross section, pipe turns or others, so as to minimise the energy that the pump has to provide to the suspension. In this regard, the type of pump, its geometry and the shear stress imposed to the suspension need to be controlled – that is, remained unchanged upon the change in scale.

4. Critical flux

This is an empirical concept described by Bacchin et al. (2006) indicating the critical permeate flow, above which irreversible fouling occurs and the device cannot be operated any longer. The critical flux corresponds to the transmembrane pressure at which the system becomes unstable due to fouling, as described in detail by Raghunath et al. (2013). Any flux value below this critical value will not impact the TMP, and fouling is reversible. In tangential flow filtration, the critical flux is influenced by the tangential flux velocity, which depends on the stepwise increase and the corresponding value of the TMP is measured as shown in Figure 5.12. The flux value at which the TMP starts to increase sharply over time reflects the appearance of fouling and is taken as the critical flux recirculation rate (Bacchin et al. 2006, Beier & Jonsson 2009). This parameter can become limiting in high-perfusion-rate operations, such as those typically used in N-1 applications.

5. Filtration loop to reactor volume ratio

This ratio represents the number of times the cells visit the filtration device and the same considerations reported earlier for the ATF unit hold true.

Considering that there is no standard equipment available on the market, the implementation of TFF units probably needs extra precautions to ensure that all the parameters described earlier are within a feasible operating range. Special care must be given to the configuration of the recirculation loop to avoid undesired shear stress contributions or pressure gradients that could impact the performance of the entire perfusion system. For the same reasons, these systems provide more flexibility for the design of specific processes compared to the ATF counterpart.

ATF   TFF

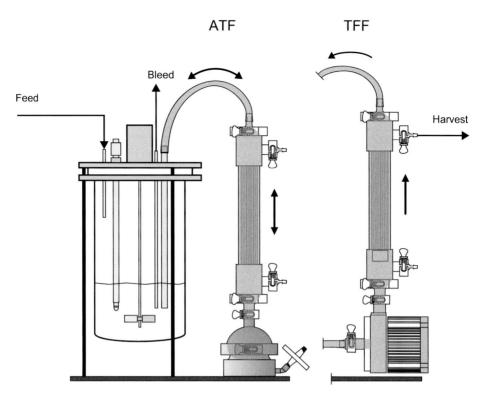

Figure 5.13 Schematic representation of laboratory ATF and TFF filtration modules: the orange color indicates the volume to be considered in the calculation of the average residence time, which is changing in the case of ATF, as described in Figure 5.10. For the TFF this includes all tubing from and to the bioreactor.

Different kinds of failures may occur when operating filtration cell retention devices, mainly due to the complexity of the suspension to be filtrated, which in perfusion bioreactors is particularly concentrated and viscous. Membrane fouling and clogging are the primary concerns, with the suspension containing small and large molecules, hydrophobic and hydrophilic compounds, cells and cell debris, and so on. On the other hand, it is also possible that due to some transport difficulties, particularly in the presence of some partial fouling, some of the target protein is also retained inside the reactor. This undesired event is referred to as sieving in the literature (Lin et al. 2017, Radoniqi et al. 2018, Wang et al. 2017) and is quantified as follows:

$$\text{Sieving} = \left(1 - \frac{C_{Harvest}}{C_{Reactor}}\right) \times 100, \qquad (5.16)$$

where $C_{Harvest}$ represents the protein concentration in the harvest and $C_{Reactor}$ represents the protein concentration inside the bioreactor. If the two are equal, then no target protein retention occurs, and the sieving effect is zero. Fouling and sieving may represent serious problems in scale-up, forcing the limitation of the operation duration or the cell density. A quantitative understanding of these processes is not available, and experimental reports are not always coherent. In general, it is believed that the alternating flow typical of ATF creates a backflush on the filtration membrane that helps to prevent fouling over longer periods of time (Bonham-Carter & Shevitz 2011, Kelly et al. 2014, Radoniqi et al. 2018). Karst et al. (2016) characterised and compared two TFF and ATF units under similar conditions using a bearingless centrifugal pump for the TFF and reported up to 50 per cent product retention when using the TFF and none with ATF. Clincke et al. (2013b) tested the limits of these two systems in wavebag cultivation systems. It appeared that the highest cell density obtained was around 210×10^6 cells/mL with the TFF system, although this density could not be sustained for longer times. Using the ATF, the maximum reached cell density was only 130×10^6 cells/mL because of the high viscosity that was limiting the liquid displacement capacity of the ATF pump. The two studies mentioned here show that both systems have different advantages and limitations that are yet to be fully understood.

In the case where fouling or sieving cannot be prevented for a given process during longer operations, it is possible to connect several filtration units to either use them in parallel or as a backup if necessary. These strategies are indeed viable, but care must be taken not to increase the contamination risk. An interesting alternative is offered by pre-sterilised and single-use devices that are described in Section 5.8.

5.7 PROCESS INTENSIFICATION

Pharmaceutical industry is experiencing unprecedented pressure from the market because of the increasing pressure on cost, the uncertainty and variability of the product demand, the market growth and regionalisation and the emergence of new classes of products. To face these challenges, alternative technologies are needed to improve the quality of the product, reduce the costs, increase the manufacturing flexibility, increase the speed to market and decrease the release time. These are crucial to answer the demand for modern biotherapeutics and prevent drug shortage or other manufacturing or supply limitations. Among many enabling technologies, some were identified by the BioPhorum Operation Group (BPOG), an international group of industrials that is collaborating to guide the technological changes in biomanufacturing. These include process technologies, automated facilities, modular and mobile facilities, in-line monitoring and real-time release, knowledge management and supply partnership management (Sawyer et al. 2017b).

In the following, we focus on process intensification, which is largely driven by continuous manufacturing solutions like perfusion. Nevertheless, perfusion alone is not an answer to the previously mentioned challenges and must be seen in a larger context. As an example, in-line monitoring techniques, discussed in Chapter 3, are crucial for controlling continuous processes but could also enable real-time batch-release procedures. These would, from a regulatory perspective, ensure the quality of the final drug substance or drug product and be valid to release a production batch. This brings along a significant number of advantages, including better process understanding, error reduction and improved product consistency by closed-loop feedback control systems. Finally, the lag time between production and actual release could be significantly reduced from weeks to days. Long lead times due to analytical testing before the release are indeed a major bottleneck (Jiang et al. 2017, Swann et al. 2017).

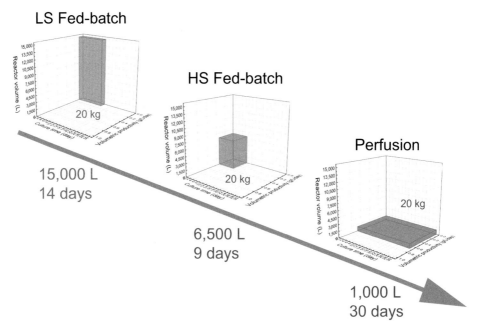

Figure 5.14 Footprint of different process types for the production of the same amount of target protein, based on the case study presented in Chapter 4, Figure 4.19. The production volumes are 15,000 L, 6,500 L and 1,000 L for traditional fed-batch, intensified fed-batch and perfusion, respectively. A relevant parameter is the run duration which is longer for the perfusion process.

To intensify existing fed-batch processes, the seed train can be intensified by introducing a perfusion bioreactor before the production bioreactor (N-1) to increase the inoculated biomass. This is done in the so-called high seeding fed-batch processes described in the example in Section 4.5.3. The objective is to increase the volumetric productivity in the production bioreactor (Bausch et al. 2018, Chen et al. 2018), which in general results from the higher seeding density leading to shorter runs with higher final titres. On the other hand, operating the production bioreactor directly in the perfusion mode can further increase the volumetric productivity, since high cell density values are maintained during the entire operation for longer times. This is well illustrated by the example of the conjugated protein described in Section 4.5.3, where it is seen that, for a fixed total production of the target protein, the bioreactor volume could be decreased from 15,000 to 6,500 L using batch-like intensified processes and even further to 1,000 L operating a perfusion bioreactor at steady state for 30 days, as shown in Figure 5.14. The comparison of the three operation modes was performed under platform conditions, which means that under optimised conditions, the specific productivity could be improved even further. Chen et al. (2018) described a hypothetical example using reasonable assumptions leading to a footprint reduction from 15,000 to 2,000 L for intensified fed-batch and to even 500 L for perfusion.

In the following, the implementation of perfusion for intensification of existing processes and end-to-end continuous integrated manufacturing is discussed through a few relevant examples. The value of single-use technologies that are often combined with these new manufacturing strategies is discussed in the next section.

Table 5.2 Freezing container type and corresponding maximum viable cell density values. © Reprinted from *Biotechnology Advances*, 36/4, Jean-Marc Bielser, Moritz Wolf, Jonathan Souquet, Hervé Broly and Massimo Morbidelli, 'Perfusion mammalian cell culture for recombinant protein manufacturing: A critical review', 2018, with permission from Elsevier.

Reference	Freezing container	Maximum Cell density / (10^6 cells/mL)
Heidemann et al. (2002)	50 or 100 mL cryobags	20 or 40 respectively
Tao et al. (2011)	5 mL cryotubes	90–100
Seth et al. (2013)	150 mL cryobags	70
Wright et al. (2015)	4.5 mL cryotubes	100

5.7.1 Process Intensification Using Perfusion

There are multiple ways to integrate perfusion to support process intensification in general. As discussed by Bielser et al. (2018), these range from high-density cell banks to the optimisation of the N-1 bioreactor in front of the production bioreactor and, finally, to the actual replacement of the fed-batch production bioreactor with a continuous perfusion one.

Table 5.2 summarises some literature references describing cell banks for high cell density inoculation. Here, perfusion technology is used as a tool to concentrate a culture with the cells in the exponential growth state. They are then frozen in either standard cryotubes (a few mL) or larger cryobags (50–150 mL). Using these concentrates, the seed train prior to a production bioreactor can be significantly shortened. From a manufacturing perspective, this is important not only because of the production time reduction, per se, but also because this allows a higher degree of flexibility at the level of production planning. It is also worth noting that, using cryobags, all operations can be performed in closed systems (no need for a laminar flow hood), which reduces significantly any contamination risk.

As already mentioned, another application of perfusion is at the level of the N-1 bioreactor located in front of the fed-batch production unit (Chen et al. 2018, Jordan et al. 2018, Pohlscheidt et al. 2013, Yang et al. 2014). In this case, medium is perfused to increase the cell density as much as possible while keeping the cells in an exponential growth phase before transferring them to the production bioreactor. The bleed outlet stream is not necessary in this case since the goal is not to reach steady-state conditions but simply to maximise viable cell density to seed at higher concentrations than in traditional fed-batch operation. This leads to shorter production runs and, consequently, lower occupancy of the production bioreactor. This again offers significant advantages in terms of planning, flexibility and, therefore, productivity, with the additional advantage that this technology can be implemented in existing facilities, originally designed for batch operation, including production bioreactors larger than 10,000 L.

Perfusion can also be used only in the initial phase of a fed-batch production. That is referred to as a hybrid process. Hiller et al. (2017), for example, describe a process in which perfusion was applied during the first 4 days only. The maximum viable cell density was increased to 60 or 80×10^6 cells/mL and the productivity increased up to 2.5-fold, depending on the clone, with respect to classical fed-batch operation. This example shows that the combination of batch-type processes with perfusion can often lead to improved performance. Perfusion could, for example, also be applied at the end of a fed-batch process to remove excessive toxic waste products, maintain a higher viability for a longer period, and even clarify the culture and eliminate standard centrifugation and depth filtration steps (Bonham-Carter 2018).

The forementioned examples show only a few of the options that one can think of to couple continuous and traditional fed-batch operation to obtain intensified processes with superior performance. Whenever possible, these are valuable options and should be pursued. However, the implementation feasibility will depend on the constraints resulting from the existing plant. To upgrade a production or N-1 bioreactor to be operated under perfusion conditions, a cell retention device and the underlying pumps for recirculation must be added. This means that enough space at the right location must be available. The media production capacity must also either already exist or else be created for that purpose. Finally, automation to control the perfusion rate is also required. The upgrade of an existing production suite is a significant investment, but considering the potential advantages demonstrated extensively in the literature, it can be an interesting option to consider.

Using a single perfusion bioreactor at the production stage is the ultimate implementation step for this technology. The complexity of the system is increased because it requires an additional control on the bleed stream to achieve steady state. The medium volumes are, compared to N-1 runs, much more important, and comparable harvest volumes need to be treated by the downstream purification units. In order to avoid storing huge volumes of harvest, the best option is therefore to operate also the capture and purification units in continuous mode. This type of operation is described hereafter.

5.7.2 End-to-End Continuous Integrated Operation

Continuous integrated protein manufacturing refers to a process where the upstream continuous bioreactor is coupled with a continuous downstream purification section and the two are operated as a single unit (Karst et al. 2018). In the classical case of monoclonal antibody production, this translates into a rather general production platform, where the perfusion bioreactor is followed by protein A capture and a couple of polishing steps, in addition to viral inactivation, as illustrated in Figure 5.15 (Baur et al. 2016, Pfister et al. 2018, Steinebach et al. 2016b, Vogg et al. 2018). The first successful demonstration of the integration of a perfusion bioreactor and a continuous capture step was reported by Warikoo et al. (2012) for

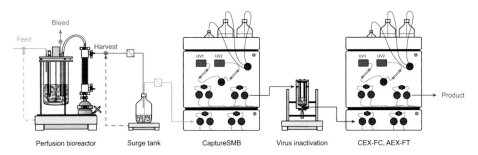

Figure 5.15 Continuous end-to-end process schematic, typical for monoclonal antibody production, including a perfusion bioreactor connected to a capture two-column simulated moving bed unit and then to viral inactivation and two polishing steps.

the production of a mAb and a recombinant human enzyme, which correspond to a stable and a complex and unstable protein, respectively. Besides any other engineering consideration, this demonstrated for the first time the dramatic reduction of the production unit footprint because of the elimination of intermediate hold tanks. This result probably marked the very beginning of continuous manufacturing in biopharmaceutical industries, which most scientists in this area now agree to have very attractive perspectives.

One crucial turning point in the development of this technology will be the massive introduction of automation and digitalisation, based on local sensors connected to supervisory control and data acquisition (SCADA) units which, properly supplemented with mathematical models, can run the unit, rejecting disturbances and keeping optimal performance, as discussed in Section 6.7. Each unit operation can be operated at its best condition, while assuring proper communication between the different units. This would have significant impact not only on the overall production costs but even more importantly on the product quality and safety, with eventual benefits to the patients.

A few demonstrations of end-to-end integrated processes, which also included viral inactivation and polishing steps, have been reported in the literature. Godawat et al. (2015) described an end-to-end process that included perfusion, continuous capture and post-capture downstream processing. They demonstrated over a longer period of time (30 days) that such a process could deliver consistent productivity and product quality. The process included a perfusion bioreactor and two periodic counter-current chromatography systems for capture and polishing, with a viral inactivation hold step in between. In terms of performance, the volumetric productivity was increased tenfold and sixfold, compared to batch operation, for upstream ($g/L_{Reactor}/day$) and downstream ($g/L_{Resin}/day$), respectively. The column size of the chromatographic steps was reduced 20-fold, and the buffer consumption reduced by 20 per cent.

A lab-scale integrated process for the production of a mAb was reported by Steinebach et al. (2017). The process included a perfusion bioreactor, a two-column continuous capture step, a two-column continuous (bind and elute) polishing step, viral inactivation and a final batch flow-through polishing step. A constant product output –in terms of concentration, aggregate and fragment content but also charge isoform and glycan distribution – was demonstrated. The contribution of each downstream step in reducing the impurity content in terms of HCP, aggregates, fragments, DNA, and protein A leached in the capture step was discussed.

An alternative operation mode has been introduced by Coffman et al. (2017) by using a short-term (<15 days) non-steady state perfusion with low perfusion rates (0.3 RV/day) coupled with an online UPLC in order to monitor product quality and titre. The perfusion bioreactor is integrated with two batch chromatographic steps, a continuous virus inactivation step, and avoids in-process pooling. In this set-up and the product is stored after the second chromatographic step for the duration of the complete batch (<15 days) and then pooled and batched through a virus reduction filter and ultrafiltration diafilatration (UFDF) towards the bulk drug substance.

Arnold et al. (2018) demonstrated the feasibility of end-to-end continuous manufacturing at pilot scale for the production of a recombinant mAb. Compared to the reference fed-batch process, comparable levels of aggregates, host cell proteins and DNA content could be achieved. They used the BioSolve software to compare the productivity of both processes and demonstrated that a fed-batch facility with $4 \times 12{,}500$ L stainless-steel bioreactors could be replaced by $5 \times 2{,}000$ L single-use perfusion bioreactors, each including its respective batch and continuous purification train, with an operating cost of goods reduction of 15 per cent.

The implementation of quality control as indicated by the regulatory authorities requires the definition of a batch – that is, a finite portion of product to which certain measured quality attributes refer. In the case of continuous manufacturing, this is not an obvious concept as it is for traditional batch operations. In general, a batch in continuous processing can be defined as the amount produced in a fixed interval of time, suitably defined in agreement with the regulatory guidelines (Allison et al. 2015). Failure management can also benefit from diverse batch definitions. In the case where, for example, a batch is defined at regular intervals in a continuous process, the product is secured on a regular basis. A contamination issue that occurs in the middle of a run would, for example, not impact the fractions that were released during the preceding portion of it.

Thus summarising, end-to-end continuous manufacturing is basically combining different enabling technologies such as continuous upstream and downstream, single-use equipment, in-line monitoring, automation and supervisory systems. One obvious benefit of an end-to-end continuous process is the removal of all the hold tanks present in batch operations, thus significantly reducing the production suite footprint. The bioreactor size is decreased because of process intensification, and the continuous downstream process allows the implementation of more efficient counter-current chromatographic processes (Pfister et al. 2018). Together, these technologies can fit in a modular production suite that can be easily parallelised and operated in a so-called ballroom design, which is defined as a manufacturing area with no fixed equipment and minimal space segregation. A further benefit could be obtained by confining the different units in closed systems. In this case, the entire zone could be unclassified, and the cost of the manufacturing space would be dramatically reduced because of common operating costs such as heating, ventilation, air conditioning and, finally, as already mentioned, the room classification (Kratzer et al. 2017, Wolton & Rayner 2014). Continuous end-to-end manufacturing is a perfect candidate to implement this concept, especially combined with single-use technologies that are addressed in the next section.

5.8 SINGLE-USE TECHNOLOGY

Single-use equipment is very popular in biomanufacturing for a number of different applications. For cell culture, it has been used for many years at the laboratory scale, while larger scale equipment has been developed more recently and is rapidly gaining acceptance (Shukla & Gottschalk 2013, Shukla & Thömmes 2010). Indeed, it offers distinct advantages which are regarded as crucial for future manufacturing processes. These include lower production costs, higher flexibility, decreased footprint and its being ideal to cope with market regionalisation (Chen et al. 2018, Sawyer et al. 2017b). Single-use equipment allows the reduction of the capital cost investment for a new plant construction and commissioning. Costs related to cleaning, validation and utilities are reduced, and rapid changeover and reduction of cross-contaminations make this equipment ideal for operation in a multiproduct site (Shukla & Gottschalk 2013).

Stainless steel holder

Control unit

Figure 5.16 A 2,000 L Mobius single-use bioreactor system. A stainless-steel holder is necessary to contain the pressure inside the single-use bag when filled with liquid. A central unit monitors different signals and controls agitation, aeration, temperature, pH and the different pumps. Courtesy of EMD Millipore Corporation.

Single-use bag

A single-use bioreactor must, by definition, be made of an FDA-approved plastic that is qualified for the manufacturing of drug products (e.g., polyethylene, ethylene-vinylacetate, polycarbonate, polystyrene) and can be used only once (Löffelholz et al. 2014, Mokuolu 2018). They exist at different scales and up to 2,000 L, as the one shown in Figure 5.16. It has been shown that single-use bioreactors can perform equally well as glass or stainless-steel bioreactors (Smelko et al. 2011). Nevertheless, there are some limitations that must be overcome to ensure proper utilisation of these equipment. The most critical are probably the leachable/extractable, potentially toxic or inhibitory compounds that these materials tend to release into the cell culture. Many studies report this problem, and the impact on cell culture growth has even been demonstrated in some cases (Dorival-García & Bones 2017, Hammond et al. 2014, Kelly et al. 2016). A second important limitation factor is the supply chain. There is, in fact, only a limited number of vendors on the market, and the quality and security of the supply is a high risk if it is not under control. To mitigate these risks, equipment standardisation and well-established relationships with vendors play an important role (Barbaroux et al. 2014, Shukla & Gottschalk 2013). On the other hand, it is also expected that if the global tendency of the industry is to move towards single-use equipment, the vendors will be in a more comfortable position to secure the supply in a more stable and sustainable manner (Kolwyck et al. 2017).

Other limitations for single-use bioreactors include the scale (currently limited to about 2,000 L), and the high cost of disposable items and waste disposal. Nevertheless, the scale limitation is in line with the plant footprint reduction and the resulting increase of flexibility, which constitutes one of the main advantages of this technology.

Figure 5.17 ATF units at different scales: the three units on the right are stainless steel and must be autoclaved before use, while the three on the left are single-use cell retention devices that include the pump head. XCell™ ATF Cell Retention Devices from Repligen. Photo provided by Repligen Corporation, all rights reserved

We have already mentioned that TFF and ATF filtration are available on the market as single-use devices for large-scale manufacturing. They offer similar advantages as single-use bioreactors in the sense that they are pre-sterilised and ready to use. Of course, the filter itself is anyway single-use, but in this case, the entire set-up, including pump head and tubing, is single-use. In the case of the non-single-use ATF system, the filtering membrane is first assembled with the pump head and the connection lines, and then the whole set-up is autoclaved together, which may be cumbersome due to the size and weight of these units. Having access to a ready-to-use system with the pump head, filter and all connection lines pre-sterilised reduces contamination risks and operational constraints, in addition to eliminating the need for a large autoclave. Both devices the stainless-steel and single-use ATF, are illustrated in Figure 5.17.

Besides bioreactors and cell retention devices, a wide range of single-use equipment exist and include media storage and preparation bags, sterile connectors, single-use probes, filters, pre-packed columns for downstream application and many others. In principle, each of these can be used to realise a fully single-use end-to-end integrated process.

5.9 ECONOMICAL DRIVERS

It has been mentioned several times that continuous bioreactors may have a beneficial effect over investment and operation costs compared to fed-batch units in different manufacturing scenarios. Indeed, this issue has been treated and analysed in the literature, with a number of valuable contributions coming from the industry. Particularly relevant here are the efforts of the already mentioned BPOG, who has generated a number of reports that discuss different scenarios of the future of biomanufacturing (available online at www.biophorum.com). The first conclusion is that there will be several types of manufacturing facilities depending mostly, but not entirely, on the product demand. Large-scale stainless-steel facilities for fed-batch will probably continue to be competitive for very large production volumes. Perfusion in single use is rather expected to be favoured in the case of medium throughput production of different molecules. This corresponds to a facility that can be easily reconfigured and to suites scaled across or, in other words, copied and parallelised (Sawyer et al. 2017b).

Of course, as mentioned before, the comparison between fed-batch and perfusion technologies depends on the specific industrial application under examination and particularly on the requested production capacity. However, the latter is not always well defined for biopharmaceuticals, or at least much less than for other products on the market. The development time lines of a successful drug can, in fact, take more than a decade, and the failure risk is extremely high. The clinical outcome during the clinical phases and even during the product lifetime play a significant role on the demand uncertainty.

As an example, immuno-oncology is a characteristic treatment field of biopharmaceuticals in which molecules are used to activate the immune system against a specific cancer (Hoos 2016). These drugs are highly promising not only because of their efficiency but also because they target only the cancer cells and do not harm the entire body as is typically the case for small molecule chemotherapies (Antonia et al. 2014, Chen 2013, Fisher et al. 2018, Kohrt et al. 2016, Lameris et al. 2014, Solomon & Garrido-Laguna 2018). Therefore, the demand for these drugs increases continuously, and their prices, which are also related to political decisions, are continuously under pressure (Howard et al. 2015, Sawyer et al. 2017b). All of this provides the motivation to provide ever-increasing amounts of these drugs to people at lower prices but without compromising efficacy and safety.

A variety of models were proposed to understand the economic impact of perfusion and, more generally, continuous processing. Pollock et al. (2013) developed a model to compare traditional fed-batch with perfusion in different industrial scenarios. In this case study, not only the production bioreactor but also the related purification steps were considered. Two perfusion processes were imagined: a first- and a second-generation one using a spin filter and an ATF as cell retention devices, respectively, with the second allowing for much higher cell density values. Of course, this kind of modelling requires many assumptions, but the conclusion was that using an ATF perfusion strategy, and assuming a fivefold increase in maximum cell density compared to fed-batch, the cost of goods (COG) savings could be up to 20 per cent. The model includes investment and consumable costs, and the COG calculations were made for different yearly productions (e.g. 100, 500 and 1,000 kg/year). With this model, considering only costs of goods, the ATF scenario is always beneficial in terms of cost. In the same study, the structure of the COG was also investigated. Without any surprise, the cost of bags and media came as the major contributors to the overall production cost.

Walther et al. (2015) developed a complex model for integrated continuous biomanufacturing with a dynamic scenario in which multiple products (stable and unstable proteins) with different annual demands are launched over 10 years. The model uses processes-economic modelling tools, Monte Carlo simulations and risk-based net present value (NPV) analysis. The overall learning for the investigated cases is that integrated continuous manufacturing offers significant financial advantages, in terms of both capital and operational costs.

Another study came to a different conclusion. Klutz et al. (2016) estimated the COGs for fed-batch and perfusion, including batch and continuous downstream, respectively. The result was that for the upstream part, the cost of goods of perfusion was systematically higher than for fed-batch. For downstream instead, continuous operation was always cheaper than batch-mode purification. Therefore, Klutz et al. (2016) suggested a hybrid process with fed-batch cell culture followed by continuous purification. This is indeed a viable strategy that can be particularly interesting in the case of existing facilities that encounter capacity limitations on the downstream side.

A relevant aspect when considering production costs refers to cell culture media, which accounts for a significant portion of the cost of goods and usually offer quite some room for optimisation (Langer 2011, Xu et al. 2017b). Media management for perfusion is not a trivial question and has already been addressed in Section 2.5. In principle, each manufacturer can choose between commercial and proprietary media. If commercial media are used, the value of the minimum cell-specific perfusion rate ($CSPR_{min}$) for the used cell line is fixed, and this defines the amount and composition of media needed with no room for further optimisation. With proprietary media, each compound can potentially be balanced to the minimum required level for a targeted CSPR. Ideally, in fact, all the compounds fed to the bioreactor should be consumed and transformed in either energy, biomass or target product. The harvest would, in this case, only contain water, metabolic waste products and the produced protein. It is obviously difficult to adjust all media components so precisely, especially as some of them like salts and pluronic are not meant to be consumed. Nevertheless, there is significant room for the cost optimisation of media composition, particularly when considering that the different components have quite different costs. Accordingly, in media development, one should always consider the price of each compound and contrast this with the benefit it brings to the culture. This work must be done early in the process development since it can lead to significant financial benefits.

In industrial practice, it should be considered that the cost of media is not only driven by the raw material value but also by the cost of preparation and storage. Consider the case where a 1,000 L bioreactor is operated at one reactor volume per day for 30 days. This means that 30,000 L of media must be prepared during such 30 days, and the question arises about what the most economic strategy is. If the entire batch of 30,000 L is prepared in advance, then huge storage tanks are necessary, which is obviously not in line with the plant footprint reduction that is envisaged for future technologies. Alternatively, one could prepare a defined volume every day or second day, but this increases operational burden and still requires some storage capacities. Other scenarios include online dilution of a concentrated feed with some buffers and online powder dilution. This is clearly not a trivial problem and further justifies the efforts that were discussed in Section 2.5 to decrease as much as possible the media consumption, which means the $CSPR_{min}$ value (McCoy et al. 2015, Xu et al. 2017b).

Besides economics, it is also important to evaluate the impact that different manufacturing technologies have on the environment. Bunnak et al. (2016) performed a life-cycle assessment (LCA) of the bioreactor operating mode by quantifying its environmental impact in terms of water consumption, energy requirements and solid waste generation. It is found that perfusion has a larger environmental impact than fed-batch, with respect to all parameters considered: 35 per cent more water, 17 per cent more energy and 17 per cent more CO_2 emission. However, the outcome of this study depends significantly on the selected downstream strategy. The perfusion harvest in this study was assumed to be pooled for four days before being further processed in batch-mode downstream purification. When this pool duration was increased to eight days, the two bioreactor operating modes were found to exhibit the same water and energy consumption. These results indicate how the environmental footprint can be sensitive to different choices about manufacturing technologies.

In conclusion, it is worth mentioning that a better understanding of the economic, financial and environmental aspects of biomanufacturing is quite relevant to guide the development of innovative technologies including those based on continuous operation. This is an essential step towards new manufacturing strategies that can bring to the market significantly safer and cheaper therapeutic proteins.

5.10 CONCLUSION

In this chapter different aspects of the scale-up of perfusion cell cultures from the laboratory to the clinical and commercial scales are discussed. The three most relevant scale-up parameters related to aeration, mixing and shear stress are discussed in detail. Scale-up characterisation and validation is then analysed with close reference to both the essential components of a perfusion system: the stirred bioreactor and the cell retention device.

Next, we analyse the role that perfusion can play in the intensification of large-scale production facilities. This is seen in the context of existing facilities based on batch-type technologies as well as with reference to integrated continuous end-to-end units, which indeed provide a fascinating alternative and a paradigm change with respect to the current biomanufacturing technologies.

A lot of innovation is currently ongoing in biomanufacturing, and most of it is ready or getting ready to be transferred to the clinical and commercial scales. This includes continuous integrated manufacturing, single-use equipment, online sensors, supervisory control and digitalisation. All of these could be combined in a single, end-to-end integrated and highly automated manufacturing platform (Fisher et al. 2018). This would probably include perfusion bioreactors between 200 and 1,000 L, which could be parallelised if needed, with high flexibility for demand fluctuations or multiproduct productions, always keeping a high level of quality control. Closed operation would minimise contamination risks and could even lead to declassified manufacturing areas, with substantially lower investment costs. By decreasing the footprint, cost and complexity of the biomanufacturing units, market regionalisation could be addressed more easily by spreading the manufacturing capacities across the world, and avoid centralised production strategies, as it often happens today (Moyle 2017).

6

Mechanistic and Statistical Modelling of Bioprocesses

This chapter provides a general introduction for the application of mechanistic and statistic models in bioreactor process development and optimisation. We first introduce the equations governing the behaviour of batch and continuous stirred tank reactors (CSTR) for a given chemical system. After presenting the criteria to generalise these concepts to biological systems and bioprocesses, we discuss the implementation of mechanistic models for the simulation and control of the bioreactor performance with particular emphasis on product quality attributes, such as N-linked glycosylation. In addition, we describe the implementation of statistical and hybrid models and their application in process development and reactor optimisation. Lastly, we compare the use of the various modelling techniques for process monitoring and control.

6.1 ROLE OF MATHEMATICAL MODELS IN UPSTREAM PROCESSES

In the previous chapters, we have introduced and discussed the most relevant concepts and hardware requirements in order to design, develop, optimise, and scale up perfusion cell culture processes. In this chapter, we discuss the use of mathematical modelling for upstream processes. Mathematical models remain the primary tool for the quantitative understanding of processes, and their use can fundamentally improve our confidence in the way we design and operate them. As such, models provide a solid basis for changing from the current state-of-the-art recipe production towards real-time online monitoring and closed-loop production control, as well as towards a better understanding of the sources of variability in processes and products (Rathore 2009, 2014, Undey et al. 2011). The applications of numerical simulations are manifold. Models can be used at the process design and development level by helping the search for better operating conditions (Karst et al. 2017b), the design of an experimental campaign intended to reach a specific objective (Brühlmann et al. 2017b, Narayanan et al. 2019c, Sokolov et al. 2018) and support the operation and control of a running unit (Steinebach et al. 2016a). The latter can be accomplished by introducing elements of forward control action or, in a more structured approach, as part of model predictive control strategies (Morari & Zafiriou 1989). Mathematical models can also be extremely valuable for the analysis of the potential of a given technology, or even at the level of process characterisation and validation for regulatory purposes (Baur et al. 2015). Models also represent a core component of several initiatives launched by regulatory agencies, such as process analytical technology (PAT) and the quality by design (QbD).

Despite recent developments in understanding the cell machinery and its inter-actions with the culture environment, process modelling in the biopharmaceutical industry is still at its infancy compared to other fields, given the enormous complex-ity of living cells (Jagschies et al. 2018). Currently, bioprocess modelling is used to simulate bioreactors at the macroscopic scale and specific biological processes at the cell level to tune the quality characteristics of the final product. Bioreactor models are based on the description of the kinetics of the chemical reactions, the mass transport among the different phases and the fluid-dynamic behaviour of the suspension inside the bioreactor. From the cell biology side, cellular processes have to be described to a suitable level of detail – such as growth, metabolism, protein production and modification pathways on a single-cell or cell-population level. These descriptions can be based on first principles, like mass or energy balances, thus leading to mech-anistic models, or the elaboration of available historical data to draw correlations of statistical nature (Kiparissides et al. 2011, Kontoravdi et al. 2010). Both these types of models not only play a significant role in process understanding but may also be useful for the implementation of soft sensors and process control (Carrondo et al. 2012, Luttmann et al. 2012). For example, the spectra collected by Raman or NIR sensors in a bioreactor need to be processed and correlated to certain observed variables in order to build predictive models that can be used to implement such sensors for process monitoring, optimisation and control.

In the next sections, we are going to introduce the basic concepts for the mod-elling of chemical systems. We will then provide the criteria and assumptions for applying this framework to biological systems with respect to different bioreactor operating modes.

6.2 MODELLING OF CHEMICAL SYSTEMS

Let us introduce the concept of reactor modelling in very general terms in order to understand which kind of knowledge is needed to build such models and which kind of information about the reactor behaviour we can extract from them. In particular, we refer to stirred tank reactors, where we assume ideal mixing conditions so that the reactor content is uniform (i.e., there is no difference in composition or temperature between any two points inside the reactor). The mass balance in such a reactor can be written in the following general form (Fogler 2008):

$$\frac{dN_i}{dt} = F_i^{in} - F_i + G_i, \tag{6.1}$$

where $\frac{dN_i}{dt}$ is the rate of accumulation of species i inside the reactor, N_i is the number of moles of the species i inside the reactor, F_i^{in} is the molar flowrate entering the reactor, F_i is the molar flowrate leaving the reactor, and G_i describes the rate of pro-duction of species i through any chemical or biological transformation. Introducing c_i, the molar concentration of species i uniform inside the reactor, and thus also characterising the outlet stream, we can use a simplified form of Equation (6.1):

$$\frac{d(c_i V_R)}{dt} = Q_{in} c_{i,in} - Q_i c_i + r_i V_R, \tag{6.2}$$

with Q representing the volumetric flowrate and r_i representing the rate of production of species i, defined by

$$r_i = \frac{\text{moles of } i \text{ produced}}{\text{time} \times \text{volume}}. \tag{6.3}$$

In addition, as we are dealing with molar species, we have to consider the stoichiometry of any given reaction – which, for the simple case of the synthesis of water out of oxygen and hydrogen, is given by

$$2H_2 + O_2 \rightarrow 2H_2O$$

$$v_A A + v_B B \rightarrow v_c C,$$

where v_i represents the stoichiometry coefficient for species i, and it is negative if the species is consumed in the reaction and positive otherwise. Due to stoichiometry, the ratio of the rate of production for each species i and the corresponding stoichiometric coefficient v_i is the same

$$\frac{r_i}{v_i} = \frac{r_{H_2}}{-2} = \frac{r_{O_2}}{-1} = \frac{r_{H_2O}}{2} = R, \tag{6.4}$$

where R represents the overall rate of the reaction. In the case that several reactions take place at the same time, the overall rate of production of a given species i, r_i is given by the sum of the contributions of all different reactions, as follows:

$$r_i = \sum_{j=1}^{N_R} v_{i,j} R_j \qquad i = 1, 2, \ldots, N_C \tag{6.5}$$

where N_R represents the number of reactions, N_C the number of species, $v_{i,j}$ the stoichiometric coefficient of species i in reaction j, and R_j the rate of reaction j.

In order to define a mechanistic model for the kinetics of a given reaction, we can proceed as follows.

For a simple reaction describing the irreversible transformation of A into B,

$$A \xrightarrow{k} B,$$

we can derive the following kinetic model, assuming first order with respect to the reactant:

$$r = kc_A, \tag{6.6}$$

where k represents the reaction rate constant and typically exhibits a temperature dependence, as described by the Arrhenius law. If we now consider the slightly more complex case of the reversible reaction between a protein enzyme E and a metal cofactor M to form the activated metal enzyme complex EM and its dissociation:

$$E + M \xrightarrow{k_{forward}} EM$$

$$EM \xrightarrow{k_{backward}} E + M,$$

we can derive an equation describing the reaction kinetics by first introducing the equilibrium constant $K_{d,M} = \frac{k_{forward}}{k_{backward}}$ representing the equilibrium conditions for the activated metal enzyme complex:

$$E + M \underset{}{\overset{K_{d,M}}{\rightleftharpoons}} EM.$$

Next, we introduce the following expressions of the relevant reaction rates:

$$r_{forward} = k_{forward} c_E c_M \qquad (6.7)$$

$$r_{backward} = k_{backward} c_{EM} \qquad (6.8)$$

$$R = r_{forward} - r_{backward} \qquad (6.9)$$

$$R = k_{forward} c_E c_M - \frac{k_{forward}}{K_{d,M}} c_{EM}, \qquad (6.10)$$

where R represents the overall net reaction rate, which at equilibrium conditions is zero, thus leading to

$$k_{forward} c_E c_M = k_{backward} c_{EM}, \qquad (6.11)$$

which leads to the classical definition of the reaction equilibrium constant for the case of association of the enzyme and the metal:

$$\frac{k_{forward}}{k_{backward}} = K_{d,M} = \frac{c_E c_M}{c_{EM}}. \qquad (6.12)$$

The description of the reaction mechanism and the derivation of the corresponding expression for the reaction rate becomes more complicated for larger systems, but it can still be accomplished using the procedures typical of chemical reaction engineering (Fogler 2008, Froment et al. 1990).

6.3 MODELLING OF BIOLOGICAL SYSTEMS

The concepts developed above for the description of chemical systems can be applied to biological processes, since in principle any fermentation can be considered as the transformation of a given mixture of reactants (metabolites) into a mixture of products (biomass, proteins, lipids, sugar and other organic molecules). In principle, two categories of mechanistic models for cell culture modelling are considered: structured and unstructured (Bibila & Robinson 1995). While unstructured models neglect the inner structure of the cell, the structured ones try to account for it by considering the cell as divided in distinct compartments, where different biochemical reactions are taking place (Kontoravdi et al. 2010, Pörtner & Schäfer 1996). It should be noted that even the most detailed and biologically consistent simulation of single-cell behaviour often fails to describe the collective behaviour of an entire population of cells (Sidoli et al. 2006). Unstructured models, on the other hand, are developed to provide a first description of the entire population and, representing an empirical basis for applying control, optimisation and process development techniques (Jang & Barford 2000). Unstructured models rely on extracellular culture variables, which are the only ones typically monitored during cultivations. This translates in a reliability domain that is usually limited to the range of experimental conditions from which they have been derived, which is often small compared to the large space of process design and optimisation. However, Pörtner & Schäfer (1996) indicated that certain cellular characteristics, such as the growth of different cell lines, follows similar kinetic rules, irrespective of the cultivation mode. Here, we present the basic concepts for unstructured cell culture modelling in bioprocessing, which is currently the most widely used approach.

The main variable for fermentation processes is the total viable cell density (or biomass) within the system. Living cells reproduce through mitosis – i.e., one organism gives rise to two exact copies of itself. If the environmental conditions are favorable, the number of cells grows exponentially in time with a specific growth rate μ [time^{-1}]. For the viable cell density X_V, the corresponding mass balance takes the form:

$$\frac{dX_V}{dt} = \mu X_V. \tag{6.13}$$

The growth rate μ is usually described as a function of the extracellular environment. For example, growth can be inhibited by reducing the temperature of the system (Wolf et al. 2019a) and largely depends on the concentration of metabolites and growth inhibitors. This effect can be modelled by considering additional terms in the calculation of the growth rate (Sou et al. 2017):

$$\mu = \mu_{max} f_{lim} f_{inh}, \tag{6.14}$$

where μ_{max} is the maximum cell growth rate of the cells at optimal conditions (i.e., maximal nutrient availability and no inhibitions), while f_{lim} is the term indicating nutrient limitation, which is usually modelled as Monod equation with respect to the metabolites of interests $c_{1,\dots,n}$:

$$f_{lim} = \frac{c_1}{K_1 + c_1} \frac{c_2}{K_2 + c_2} \cdots \frac{c_n}{K_n + c_n}, \tag{6.15}$$

where K_n is the Monod constant for the limiting metabolite c_n. For mammalian cell cultures, the medium components that represent the term c_n are usually glucose, particular amino acids and growth factors. The term f_{inh} can be written as

$$f_{inh} = \frac{K_1}{K_1 + I_1} \frac{K_2}{K_2 + I_2} \cdots \frac{K_n}{K_n + I_n}, \tag{6.16}$$

where I_n indicates the concentration of a given growth inhibitor n, such as lactic acid or ammonia. The concentrations of metabolites and inhibitors follows a general mass balance:

$$\frac{dc_i}{dt} = q_i X_V + Q_{in} c_{i,in} - Q_{out} c_i, \tag{6.17}$$

where q_i indicates the specific production rate of the species i and the terms Q_{in} and Q_{out} are the volumetric flowrates that account for feeding or removal of nutrients.

Besides growth, it is usually worth considering a death term for cells, in order to be able to describe the stationary phase of a cultivation. This is usually done by modifying the mass balance for X_V as follows:

$$\frac{dX_V}{dt} = \mu X_V - \mu_d X_V, \tag{6.18}$$

where μ_d is the death rate of the cells and can be again expressed as a function of the extracellular environment. For example, Villiger et al. (2016a) modelled the death rate of CHO cells as a function of extracellular ammonia according to the following equation:

$$\mu_d = \frac{\mu_d^{max}}{1 + \left(\frac{k_{d,amm}}{c_{amm}}\right)^2},$$

(6.19)

where μ_d^{max} is the maximum possible cell death rate and $k_{d,amm}$ a constant to relate the concentration of ammonia c_{amm} to the death rate. Lastly, antibody production can be expressed as

$$\frac{dc_{mAb}}{dt} = q_{mAb}X_V,$$

(6.20)

where q_{mAb} is the cell-specific productivity, which can either be assumed constant or can vary according to the environmental conditions. Usually, it is expressed as a function of the growth rate of the cells, since non-dividing cells usually show a higher productivity compared to rapidly proliferating cells, or process variables such as pH or osmolality.

Although the presented equations are generally valid for CHO cells, they can be modified to better represent the behaviour of other cell types (Tziampazis & Sambanis 1994). Moreover, as it will be discussed later, these equations can be complemented with additional ones to describe other aspects of bioprocessing such as the quality profiles of the secreted proteins. The following section describes how these equations can be incorporated in the mass balances that describe the different operation modes of the bioreactor.

6.4 BIOREACTOR OPERATION MODES

6.4.1 Batch Operation

The simplest operation mode for a stirred tank reactor – i.e., the batch mode – is based on the introduction of all reactants at the beginning of the operation, without any inlet or outlet stream (Figure 6.1) so that the general mass balance in Equation (6.2) reduces to

$$\frac{d(c_i V_R)}{dt} = r_i V_R,$$

(6.21)

which, by assuming constant reactor volume, V_R leads to

$$\frac{dc_i}{dt} = r_i,$$

(6.22)

with the initial conditions $c_i = c_i^0$ at time $t = 0$. In this special case of an ideal reactor system, the rate of accumulation of a species i in the reactor is equal to the rate of consumption or production of the same species.

We already indicated that the state-of-the-art operation of bioreactors used for the production of biopharmaceuticals is fed-batch. This operating mode can be modelled starting from the batch case (Section 6.4.1) by simply adding an inlet stream

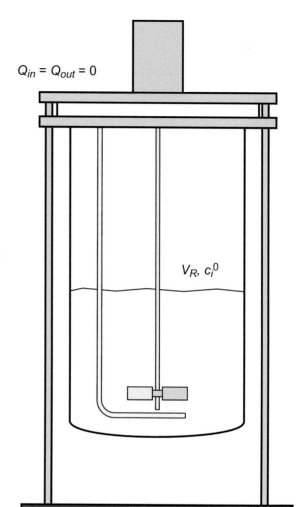

$Q_{in} = Q_{out} = 0$

V_R, c_i^0

Figure 6.1 A schematic representation of a batch reactor with constant volume, V_R and initial concentrations $c_i = c_i^0$.

with volumetric flowrate, $Q_{in}(t)$ and composition defined by the molar concentrations, $c_{i,in}$ (Figure 6.2):

$$\frac{d(c_i V_R)}{dt} = Q_{in}(t)c_{i,in} + r_i V_R(t).$$
(6.23)

Since cells are not present in the inlet stream, the corresponding mass balance reduces to

$$\frac{d(V_R X_V)}{dt} = \mu X_V V_R - \mu X_V V_R.$$
(6.24)

For oxygen and carbon dioxide, the mass balance needs to take into account the volumetric mass transfer coefficient $k_L a$ $[hr^{-1}]$, which, multiplied by the driving force, $\Delta C_{Gas} = (C_{Gas}^* - C_{Gas})$, gives the flowrate of the component coming in

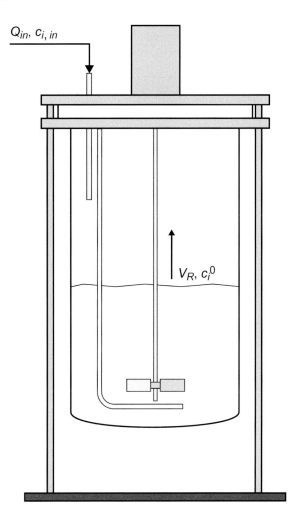

Figure 6.2 A schematic representation of a fed-batch reactor with the addition of a feed stream with volumetric flowrate, Q_{in}, and composition, $c_{(i,in)}$, with reactor volume, V_R, changing in time and initial concentrations $c_i = c_i^0$.

(out) the liquid phase, while the reaction term provides the flow of the same species consumed (produced) by the reactive processes inside the reactor:

$$\frac{dC_{gas}}{dt} = k_L a(C_{Gas}^* - C_{Gas}) + r_i V_R(t), \tag{6.25}$$

where C_{Gas}^* indicates the saturated concentration of the gaseous species of interest in the medium.

Due to the addition of the feed stream, the reactor volume changes during operation, as described by the following equation:

$$\frac{dV_R}{dt} = Q_{in}(t), \tag{6.26}$$

with the initial condition $V_R = V_R^0$ at $t = 0$.

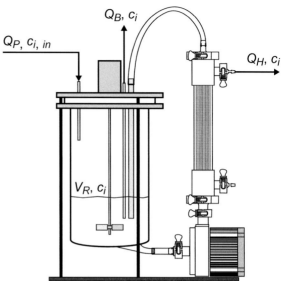

Figure 6.3 A schematic representation of a perfusion bioreactor with the perfusion stream having volumetric flowrate, Q_P, and composition, $c_{i,in}$, and two outlet streams, the bleed and the harvest, with volumetric flowrates Q_B and Q_H, respectively.

It is worth mentioning that, in the case of smaller bioreactors, the amount of volume removed through sampling, Q_{out}, is not negligible and should be taken into account in the overall mass balance as follows:

$$\frac{dV_R}{dt} = Q_{in}(t) - Q_{out}(t). \tag{6.27}$$

Note that for fed-batch systems, neither $Q_{in}(t)$ nor $Q_{out}(t)$ are continuous functions with respect to time, since sampling and feeding are usually performed at discrete time points during the cultivation.

6.4.2 Continuous Perfusion Operation

In Chapter 2, we have introduced the ideal CSTR model to simulate a stirred tank reactor operated in the perfusion mode, with the strong simplification that the cell retention device is placed inside the reactor tank itself. However, this does not usually hold true, since the filtration device is located outside the reactor and connected to it through proper piping. Accordingly, the cell retention device needs to be modelled as well. With reference to the perfusion bioreactor coupled to a tangential flow filtration (TFF) device shown in Figure 6.3, and assuming that the two outlet streams have the same composition as inside the reactor c_i, the overall mass balance and the mass balances for the individual metabolite species i can be written as follows:

$$Q_P = Q_B + Q_H \tag{6.28}$$

$$V_R \frac{dc_i}{dt} = Q_P c_{i,in} - (Q_B + Q_H)c_i + r_i V_R. \tag{6.29}$$

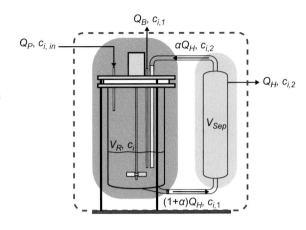

System 1   System 2

$Q_B, c_{i,1}$

$Q_P, c_{i,in}$

$\alpha Q_H, c_{i,2}$

$Q_H, c_{i,2}$

V_{Sep}

V_R, c_i

$(1+\alpha)Q_H, c_{i,1}$

Overall

Figure 6.4 Schematic representation of a perfusion bioreactor where the stirred tank (red) is modelled as an ideal CSTR system and the filtration device (light blue) can be modelled as a tubular ideal plug-flow reactor, PFR.

Assuming perfect cell retention (i.e., X_V in the harvest stream is equal to zero), the mass balance for the viable cell density, X_V reads

$$V_R \frac{dX_V}{dt} = -X_V Q_B + \mu X_V V_R - \mu_d X_V V_R. \tag{6.30}$$

In the metabolite mass balances (Equation (6.28)), we have assumed that the residence time in the external loop is much shorter than the time required for biological reactions to take place. Accordingly, the stream entering the loop is recycled unaltered back into the reactor, except for a portion corresponding to the harvest stream, $Q_H c_i$, which leaves the system. However, the residence time in the filtration module is, in some cases at the large scale, not negligible, and the same processes occurring inside the stirred tank are also taking place here and therefore need to be properly modelled. This can be done by considering the tank and the filtration module separately, as illustrated in Figure 6.4. The first one consists of a stirred vessel and is therefore reasonable to apply again the ideal CSTR model, as done earlier. For the second, however, we need to consider that the flow enters the module with a certain composition, $c_{i,1}$, equal to the corresponding composition c_i inside the stirred tank, which is then going to change as the flow proceeds along the hollow fibres and becomes, $c_{i,2}$ as the flow reaches the end of the filtration device and enters back into the stirred vessel. Since there is no stirring in this device, it appears reasonable to assume that the suspension progresses through the device like a liquid through a tube, which can be modelled using the ideal plug flow reactor (PFR) model (Fogler 2008, Froment et al. 1990). In a typical perfusion set-up, the three inlet and outlet stream flowrates – Q_P, Q_B and Q_H – are kept constant at the desired set-point value by the control system. The flow stream entering the filtration module is instead controlled by the pump of the TFF, and therefore, it is known and assumed equal to the harvest volumetric flowrate, Q_H, increased by a factor $(1+\alpha)$ in the following.

The overall mass balance of the coupled systems is unchanged with respect to the previous case:

$$Q_P = Q_B + Q_H, \tag{6.31}$$

which is equivalent to what one would write with reference to the stirred tank alone (red in Figure 6.4):

$$Q_P + \alpha Q_H = Q_B + (1 + \alpha)Q_H. \tag{6.32}$$

About the singe species i, the mass balances in Equation (6.28), are now changed and read as follows:

$$V_R \frac{dc_i}{dt} = Q_P c_{i,in} + \alpha Q_H c_{i,2} - (Q_B + (1 + \alpha)Q_H)c_{i,1} + r_i V_R, \tag{6.33}$$

while for the viable cell density, X_V, the mass balance for Equation (6.28) becomes

$$V_R \frac{dX_{V,1}}{dt} = X_{V,2}\alpha Q_H - X_{V,1}(Q_B + (1 + \alpha)Q_H) + \mu X_{V,1}V_R - \mu_d X_{V,1}V_R, \tag{6.34}$$

and for gaseous species like oxygen and carbon dioxide, we get

$$V_R \frac{dC_{Gas}}{dt} = k_L a(C_{Gas}^* - C_{Gas}) + \alpha Q_H c_{i,2} - (Q_B + (1 + \alpha)Q_H)C_{Gas} + r_i V_R. \tag{6.35}$$

Note that in agreement with the assumption of ideal mixing in the stirred tank, the conditions at the inlet of the filtration device are the same as those inside the reactor so that $c_{i,1} = c_i$ and $X_{V,1} = X_V$.

We now need to introduce the model of the filtration device in order to connect the outlet quantities, $c_{i,2}$ and $X_{V,2}$, to the inlet ones, $c_{i,1}$ and $X_{V,1}$ (light blue in Figure 6.4). As mentioned before, it is reasonable to model it as a tubular reactor with a distributed outlet liquid stream leaving the reactor while moving along its axis. This stream has the same composition as the liquid inside the loop, except for the viable cell density, which is equal to zero due to the filtering effect. In this frame, we obtain the following mass balances for the metabolite i:

$$\frac{d(Qc_{i,1})}{dt} = -c_{i,1}\frac{dQ}{dV} + r_i, \tag{6.36}$$

and for the viable cell density X_V:

$$\frac{d(QX_V)}{dV} = \mu X_{V,1} - \mu_d X_{V,1}, \tag{6.37}$$

where V is the axial coordinate along the the filter going from $V = 0$ to $V = V_{sep}$.

Note that Equation (6.36) can be applied also to oxygen and carbon dioxide, since in this portion of the unit, there is no separate gaseous phase present, like for example, bubbles.

The flowrate of the liquid stream inside the filter decreases along the axial coordinate going from $(1 + \alpha)Q_H$ at the inlet to αQ_H at the outlet. Assuming as a first approximation that the harvest flux remains constant along the coordinate V, we obtain the following expression for the liquid flowrate inside the external loop:

$$Q = (1 + \alpha)Q_H - \alpha Q_H \left(\frac{V}{V_{sep}}\right) \tag{6.38}$$

Integrating Equations (6.36) and (6.37), from the initial condition $c_{i,1} = c_i$ and $X_{V,1} = X_V$ at $V = 0$, the outlet conditions $c_{i,2}$ and $X_{V,2}$ are obtained at $V = V_{sep}$.

In the case of lab-scale perfusion bioreactors, external flowrates in the range of 1.0–1.5 L/min are reported for both ATF and TFF systems, while the harvest rate typically ranges between 0.7 and 1.0 RV/day. For the case of a 2 L bioreactor system, this equals 1.4–2.0 L/day (0.07–0.0014 L/min), giving α values ranging between 700 and 1,100, which means that the rate through the external loop is about 1,000 times larger than the overall perfusion rate.

The average residence time, τ_{Sep}, in the cell retention device can be computed as follows:

$$\tau_{Sep} = \frac{V_{Sep}}{(1 + \alpha)Q_H}. \tag{6.39}$$

As discussed in the previous chapter, this is a critical parameter during scale-up since the environment in the cell retention device is not controlled in terms of pH, temperature, and composition, including oxygen concentration. The latter is likely the most critical one since the only oxygen available to the cell in this device is the one dissolved in the suspension at the entrance. As a rule of thumb, the residence time in the cell retention device has been reported for ATF devices to be critical if longer than about 60 seconds (Walther et al. 2019). Based on the model, it is possible to investigate the influence of the residence time on any of the cell culture parameters, including the dissolved oxygen concentration. This is a very useful tool that could provide flexibility in the scale-up. For example, if we are able to quantify the rate of oxygen consumption per passage in the cell retention device, we do not necessarily need to satisfy the rigid constraint of keeping the residence time constant during scale-up, at least with respect to oxygen consumption.

6.5 MECHANISTIC MODELS AND THEIR IMPLEMENTATION

6.5.1 Cell Culture and Cell Metabolism

Mechanistic or *white-box* models of biological systems, such as mammalian cell cultures, involve various degrees of structure and mathematical complexity, which tend to be higher compared to the reactor models introduced earlier (Sidoli et al. 2004). Generally speaking, modelling of biological organisms is aimed at improving the biological understanding of the complex cellular mechanisms, while also providing a framework for the control and optimisation of cell cultures. In the frame of bioreactor models, the main aim is to describe the interaction between cell-related parameters, such as cell-specific growth and productivity, to the operating conditions inside the bioreactor (Dhir et al. 2000, Dowd et al. 1999).

The first step of a mechanistic modelling approach involves translating the *a priori* knowledge into the mathematical description of the problem, which corresponds to the model development step, followed by the analysis of the model (Kiparissides et al. 2011). This contains an *a priori* identifiability analysis of all parameters or, in the case of dynamic, non-linear models, a sensitivity analysis that studies how much a change in the model output can be attributed to the variation of the different input parameters. This step is typically used to improve model accuracy and to identify the set of parameters that needs to be estimated from experimental data. The reliable estimation of parameters, however, is only feasible if a sufficiently rich and diverse data set is available. Therefore, the use of optimal experimental design techniques is necessary to enable accurate parameter estimation and model validation (Kontoravdi et al. 2010). The last step requires the validation of the predictive capability of the model by comparison with a new set of independent experimental data.

In general, cell culture models necessitate a large number of differential and algebraic equations coupled with various parameters that need to be estimated. However, although we are still not in the position of developing models which couple a detailed description of the cell metabolism with that of the entire bioreactor, it is definitely possible to do so for more limited models aimed at the simulation of only specific quantities – for example, some structure characteristics of the target protein. In the following, we illustrate the power of mathematical mechanistic models, in particular, for the control and modulation of N-linked glycosylation patterns in mammalian cell culture processes (Jimenez del Val et al. 2011, Karst et al. 2017b, Villiger et al. 2016a).

6.5.2 N-Linked Glycosylation Modelling

N-linked glycosylation of therapeutic proteins is particularly well suited for the application of mathematical models. Contrary to other biological processes (e.g., metabolic and gene regulatory networks), the network of reactions and the enzymes involved in protein N-glycosylation are established, allowing the possibility for a relatively precise description of the entire system. On the other hand, the N-linked glycosylation patterns strongly affect the biological function of the protein, and therefore, their detailed understanding is crucial to optimise their therapeutic efficacy.

In eukaryotic cells, the first step of N-linked glycosylation occurs in the endoplasmic reticulum (ER). A large oligosaccharide (OS) structure (Glc_3 Man_9 $GlcNAc_2$) is transferred en bloc to a specific sequon (ASN-X-Ser) of the protein (Helenius & Aebi 2001). The site occupancy and the extent of additional modifications on the oligosaccharide chain depend on the efficiency of several subsequent enzymatic steps occurring in the secretory pathways. Factors that have an effect on the glycosylation process include the amount and localisation of various enzymes (kinases, transferases, permeases and epimerases), glycan–protein interactions (Losfeld et al. 2017), the pH in the lumina, and the flux and velocity of glycosylated proteins passing through the Golgi apparatus (Rivinoja et al. 2009, Roy 2009). Moreover, it is known that metal cofactors and the availability of nucleotide activated sugars (NS) affect the enzymatic kinetics of the oligosaccharide complex and many glycosyltransferases (Gramer et al. 2011, Kochanowski et al. 2008). Interference of these various enzymatic steps can result in differences in the glycosylation profile at a specific site (microheterogeneity). Several mathematical models have been reported, focusing mostly on glycan processing in the Golgi apparatus (Krambeck & Betenbaugh 2005, Umaña & Bailey 1997).

Recombinant monoclonal antibodies represent a special class of therapeutic proteins with respect to N-linked glycosylation. As they belong to the class of immunoglobulin G (IgG) proteins, they consistently bear N-linked glycosylation at the crystalline (Fc) region of both heavy chains with high occupancy, although they can also present other glycosylation sites in the variable region. The prevalent diantennary glycan structures are relatively simple compared to highly branched structures reported for various other glycoproteins. In the case of mAb Fc N-linked glycosylation, this allows us to consider only the microheterogeneity with a simplified network, which constitutes an ideal case for the development and application of a mathematical modelling framework in order to describe, predict and control N-linked glycosylation patterns. In a first proposal, this process cascade was described and simulated as a series of four ideally mixed compartments with the mass balances of the individual structures based on an enzyme-catalysed network (Umaña & Bailey 1997). Further expansion incorporated the addition of the fucosylation, galactosylation and sialylation steps ending up in a large network with thousands of substances and reactions (Krambeck & Betenbaugh 2005). Recent investigations were focused on dynamic models of monoclonal antibody glycosylation that link cellular metabolism via nucleotide sugar donors in order to couple the nucleotide sugar synthesis to the mechanistic glycosylation model (Jimenez del Val et al. 2011). This can be applied in a rather general way, and the dynamic model can be linked to any (of course, reasonable) unstructured or structured model, such as a simple cell culture model.

Mechanistic N-Linked Glycosylation Model In two exemplary case studies, Villiger et al. (2016b) and Karst et al. (2017b) used this model description and investigated the effect of manganese, galactose, culture pH and the ammonia concentration on the glycosylation pattern both in fed-batch and perfusion operations. In particular, they further developed the model of Jimenez del Val et al. (2011) and used it to describe the N-linked glycosylation trimming and elongation process based on the biosynthetic pathway within the Golgi shown in Figure 6.5. This model assumes that the first step of glycan processing taking place in the ER is fully completed – i.e., that every single mAb Fc region entering the Golgi has the same structure (M9). The Golgi apparatus itself is modelled as a single PFR, which includes the recycling of Golgi-resident proteins and transport equations for nucleotide sugars (Jimenez del Val et al. 2011). The mass balances of the different oligosaccharide structures OS_i; nucleotides, N_k; and nucleotide sugars, NS_k, are calculated considering the Golgi PFR model operating at steady state and the simplified glycosylation kinetic model shown in Figure 6.5, involving 43 reactions with 33 different glycan structures, as follows:

$$\frac{\delta \left[OS_i(z,t) \right]}{\delta t} = -\frac{4q}{\pi D_i^2} \frac{\delta \left[OS_i(z,t) \right]}{\delta z} - \sum_{j=1}^{NR} v_{i,j} r_j \tag{6.40}$$

$$\frac{\delta \left[N_k^{Golgi}(z,t) \right]}{\delta t} = -F_{T,k} - \frac{4q}{\pi D_i^2} \frac{\delta \left[N_k^{Golgi}(z,t) \right]}{\delta z} - \sum_{j=1}^{NR} v_{k,j} r_j \tag{6.41}$$

$$\frac{\delta \left[NS_k^{Golgi}(z,t) \right]}{\delta t} = -F_{T,k} - \frac{4q}{\pi D_i^2} \frac{\delta \left[NS_k^{Golgi}(z,t) \right]}{\delta z} - \sum_{j=1}^{NR} v_{k,j} r_j, \tag{6.42}$$

where D_i is the Golgi diameter, $v_{k,j}$ the stoichiometric coefficient, r_j the reaction rate, NR is the number of enzymatic reactions, q is the flowrate in the Golgi apparatus and

$F_{T,k}$ a term describing the flowrate of sugar precursors into the Golgi, which depends on the gradient of nucleotides and nucleotide sugars across the Golgi membrane:

$$F_{T,k}(z) = k_{T,k} \left[TP_k(z)\right] \left(\frac{\left[NS_k^{cyt}\right]}{k_{NS,k}^{cyt} + \left[NS_k^{cyt}\right]}\right) \left(\frac{\left[N_k^{cyt}\right]}{k_{N,k}^{cyt} + \left[N_k^{cyt}\right]}\right) \quad (6.43)$$

where $k_{T,k}$ represents the transport turnover rate and $[TP_k(z)]$ the concentration of the transport protein k at a Golgi position z. The transport proteins TP_k as well as the glycosyltransferases E_j are located along the Golgi with a Gaussian distribution:

$$[TP_k(z)] = [TP_k^{max}] exp \left[-\frac{1}{2}\left(\frac{z - z_k^{max}}{\omega_k}\right)^2\right] \quad (6.44)$$

$$[E_j(z)] = [E_j^{max}] exp \left[-\frac{1}{2}\left(\frac{z - z_j^{max}}{\omega_j}\right)^2\right], \quad (6.45)$$

with TP_k^{max} representing the peak concentration of a transport protein at a certain Golgi length z_k^{max} and E_j^{max} representing the peak concentration of a glycosyltransferase j at a certain Golgi position z_j^{max}.

As mentioned before, the 43 different reactions in Figure 6.5 are catalysed by seven enzymatic reactions whose kinetics is described in the following. First, $\alpha - 1, 2$ mannosidase I (ManI) and $\alpha - 1, 3 / \alpha - 1, 6$ mannosidase II (ManII) remove mannose (MAN) units from the glycan structure in the earliest part of the pathway (Tabas & Kornfeld 1979, Velasco 1993). Their kinetics were assumed to follow the well-known Michaelis–Menten kinetics:

$$r_j = \frac{k_{f,j} [E_j] [OS_i]}{K_{d,i} \left(1 + \frac{[OS_i]}{K_{d,i}} + \sum_{z=1}^{NC} \frac{[OS_z]}{K_{d,z}} + \frac{[OS_{i-1}]}{K_{d,i-1}}\right)}, \quad (6.46)$$

with $k_{f,j}$ representing the turnover rate of the reaction and $K_{d,i}$ representing the dissociation constant of the specific donor-enzyme complex. Next, Villiger et al.

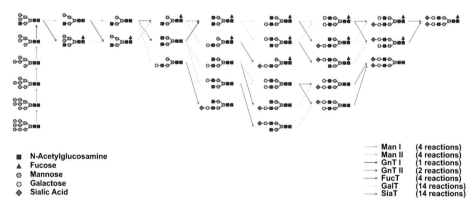

■ N-Acetylglucosamine	
▲ Fucose	
◎ Mannose	
○ Galactose	
◆ Sialic Acid	

→	Man I	(4 reactions)
⋯⋯>	Man II	(4 reactions)
→	GnT I	(1 reactions)
⋯⋯>	GnT II	(2 reactions)
→	FucT	(4 reactions)
→	GalT	(14 reactions)
→	SiaT	(14 reactions)

Figure 6.5 Simplified network describing mAb FC N-linked glycosylation used for the dynamic mechanistic models. The network consists of 33 different glycan structures (11 of them are measured), and 43 reactions catalysed by seven enzymes. © Reprinted from *Biotechnology Progress*, Thomas K. Villiger, Anaïs Roulet, Arnaud Périlleux, Matthieu Stettler, Hervé Broly, Massimo Morbidelli and Miroslav Soos, 'Controlling the time evolution of mAb N-linked glycosylation: Part I: Microbioreactor experiments', 2016, with permission from John Wiley and Sons

(2016b) considered $\alpha - 1, 3$ N-acetylglucosaminyl transferase I (GnTI) and $\alpha - 1, 6$ N-acetylglucosaminyl transferase II (GnTII), which are responsible for the addition of N-Acetylglucosamine to the glycan chain (in total, three reactions) (Bendiak & Schachter 1987). Similarly, $\beta - 1, 4$ Galactosyltransferase (GalT) catalyses the addition of galactose to the glycan chain, and is involved in 14 such elongation reactions (Bendiak & Schachter 1987, McCracken et al. 2014). The enzyme kinetics of these three glycosyltransferases are assumed to be based on sequential-order bi–bi kinetics, and their activity is highly influenced by the presence of metal co-factors (Gramer et al. 2011). Of particular interest for the enzyme kinetics is the presence of manganese ions $\left[Mn^{2+}\right]$. The metal co-factor mediates the binding between the enzyme and the nucleotide sugar, which is necessary in order to provide the required conformational changes for the binding to the N-glycan substrate. After the formation of the modified N-glycan, the glycan structure is released first, followed by the manganese ion and a phosphate group that is released as a by-product (Lairson et al. 2008). This specific type of kinetics can be expressed as follows:

$$r_j =$$

$$\frac{k_{f,j}\left[E_j\right]\left[Mn^{2+}\right]\left[NS_k^{Golgi}\right][OS_i]}{K_{d,Mn^{2+}}+K_{d,k}K_{d,i}\left(\begin{array}{c} 1+\frac{[Mn^{2+}]}{K_{d,Mn^{2+}}}+\frac{[Mn^{2+}]}{K_{d,Mn^{2+}}}\frac{\left[NS_k^{Golgi}\right]}{K_{d,k}}+\frac{[Mn^{2+}]}{K_{d,Mn^{2+}}}\frac{\left[NS_k^{Golgi}\right]}{K_{d,k}}\frac{[OS_i]}{K_{d,i}} \\ +\frac{[Mn^{2+}]}{K_{d,Mn^{2+}}}\frac{\left[NS_k^{Golgi}\right]}{K_{d,k}}\sum_z^{NC}\frac{[OS_z]}{K_{d,z}}+\frac{[OS_{i+1}]}{K_{d,i+1}}\frac{\left[N_k^{Golgi}\right]}{K_{d,Nk}}+\frac{\left[N_k^{Golgi}\right]}{K_{d,Nk}} \end{array}\right)},$$

$$(6.47)$$

with $K_{d,k}$ representing the dissociation constant of the donor-enzyme complex, $K_{d,Mn^{2+}}$ representing the dissociation constant of the manganese-enzyme complex and $K_{d,Nk}$ representing the dissociation constant of the nucleotide-enzyme complex. Another glycosyltransferase involved in the reaction scheme is $\alpha - 1, 6$ fucosyltransferase (FucT) that catalyses the transfer of a fucose molecule to the glycan chain (overall, four reactions in the network). The last enzyme of this network is the $\alpha - 1, 6$ sialyltransferase (SialT), which is involved in the addition of sialic acid to the nucleotide chain. Overall, the enzyme is involved in 14 reactions of the network. The enzyme kinetics of both transferase enzymes, FucT and SiaT, can be approached by applying a random-order bi–bi kinetic rate, expressed as follows:

$$r_j = \frac{k_{f,j}\left[E_j\right]\left[NS_k^{Golgi}\right][OS_i]}{K_{d,k}K_{d,i}\left(\begin{array}{c} 1+\frac{\left[NS_k^{Golgi}\right]}{K_{d,k}}+\frac{[OS_i]}{K_{d,i}}+\sum_{z=1}^{NC}\frac{[OS_z]}{K_{d,z}}+\frac{\left[NS_k^{Golgi}\right]}{K_{d,k}}\frac{[OS_i]}{K_{d,i}} \\ +\frac{\left[NS_k^{Golgi}\right]}{K_{d,k}}\sum_{z=1}^{NC}\frac{[OS_z]}{K_{d,z}}+\frac{\left[N_k^{Golgi}\right]}{K_{d,Nk}}\frac{[OS_{i+1}]}{K_{d,i+1}}+\frac{\left[N_k^{Golgi}\right]}{K_{d,Nk}}+\frac{[OS_{i+1}]}{K_{d,i+1}} \end{array}\right)}$$

$$(6.48)$$

Having summarised the enzyme kinetics for the N-linked glycosylation network, the next step is to evaluate the activity of the glycosylation enzymes as a function of the environmental conditions prevailing in the Golgi apparatus. Typically, the Golgi pH (pH_{Golgi}) plays the most important role with respect to the enzymes activity (Rivinoja et al. 2009). In mammalian cells, the Golgi pH, referred to as pH_{Golgi}, is slightly acidic (5.9–6.5) and well regulated in normal conditions. It is well known that extracellular ammonia and other bases that easily diffuse through the cell membranes can have a significant effect on the intracellular pH_{Golgi}, resulting in mislocalisation and reduced activity of the glycosylation enzymes. However, in most cases, it is assumed that the enzyme position across the Golgi apparatus remains unaffected, and only the enzyme activity is changed by varying intracellular pH. With respect to mammalian cell cultures, the effect of increasing ammonia concentration in the cell culture leading to larger values of the intracellular pH_{Golgi} can be expressed by the Henderson–Hasselbalch equation. The pH dependency of the activity of Golgi resident enzymes was estimated by Villiger et al. (2016c) from literature data of in vitro enzymatic assays, determining an optimal pH for the activity of Golgi-resident enzymes pH_{opt}^{Golgi} of 6.59. The turnover rate constant as well as the Henderson–Hasselbalch equation for the pH_{Golgi} can therefore be expressed as follows:

$$k_{f,j} = k_{f,j}^{max} exp \left[-\frac{1}{2} \left(\frac{pH^{Golgi} - pH_{opt}^{Golgi}}{\omega_{f,j}} \right)^2 \right] \qquad (6.49)$$

$$pH^{Golgi} = pK_A^{Golgi} + log \left(\frac{[AMM]}{N_A^{Golgi} - [AMM]} \right), \qquad (6.50)$$

with $k_{f,j}^{max}$ representing the maximum turnover rate of a specific reaction, $\omega_{f,j}$, the width of the pH activity profile of E_j, pK_A^{Golgi} the apparent pK_A value of the Golgi, $[AMM]$ the concentration of ammonia and N_A^{Golgi} the ammonia-associated Golgi pH constant. Due to the strong connection of the pH^{Golgi} and the pK_A^{Golgi}, these parameters cannot be estimated *a priori*, and therefore, Villiger et al. (2016b) provided a reasonable buffer capacity by setting the pK_A^{Golgi} equal to 7.5 at a pH value of 6.5 and used only the N_A^{Golgi} as a fitting parameter.

This detailed dynamic glycosylation model needs to be linked to a cell culture model that describes the external measured parameters, such as viable cell density, cell-specific productivity, ammonia production and intracellular nucleotide sugars and, thus, links these parameters to the one involved in the dynamic glycosylation model. Villiger et al. (2016b) and Karst et al. (2017b) proposed to use simple unstructured cell culture models to describe both N-linked glycosylation within fed-batch and perfusion bioreactor operations to overcome practical parameter identifiability and error propagation to the glycosylation model. Therefore, in both applications, the mass balances were kept simple and with the least amount of parameters to describe different operating conditions. In the following section, we focus on the connection of the dynamic N-linked glycosylation model to the different kind of operating conditions.

Fed-Batch Culture for IgG Production Villiger et al. (2016b) connected the dynamic model to describe N-linked glycosylation with a simple unstructured cell culture model used to complete the bioreactor model. The mass balances describing the culture volume and concentrations of manganese and galactose in the extracellular medium take into account perfect mixing and negligible consumption:

$$\frac{dV_R}{dt} = Q_{in} - Q_{out} \qquad (6.51)$$

$$\frac{dV_R\left[Mn^{2+}\right]}{dt} = Q_{in}\left[Mn^{2+}_{in}\right] - Q_{out}\left[Mn^{2+}\right] \tag{6.52}$$

$$\frac{dV_R\left[Gal\right]}{dt} = Q_{in}\left[Gal_{in}\right] - Q_{out}\left[Gal\right], \tag{6.53}$$

where Q_{in} is the feed volumetric flowrate, Q_{out} the flowrate out of the reactor (sampling), $\left[Mn^{2+}_{in}\right]$ and $\left[Gal_{in}\right]$ the inlet concentration of manganese and galactose, and $\left[Mn^{2+}\right]$ and $\left[Gal\right]$ the corresponding concentrations inside the reactor.

In addition, the model includes differential equations for the viable cell density X_V, the dead cell density X_d, and lysed cell population X_l as a function of the growth rate μ, death rate μ_d and lysis rate μ_l:

$$\frac{dV_R X_V}{dt} = \mu V_R X_V - \mu_d V X_V - Q_{out} X_V \tag{6.54}$$

$$\frac{dV_R X_d}{dt} = \mu_d V_R X_d - \mu_l V X_d - Q_{out} X_d \tag{6.55}$$

$$\frac{dV_R X_l}{dt} = \mu_l V_R X_d - Q_{out} X_l. \tag{6.56}$$

The growth and dead rates can be assumed to be functions of the ammonia concentration in the extracellular medium and can be described as follows:

$$\mu = \mu_{max}\left(\frac{K_{\mu,AMM}}{K_{\mu,AMM} + [AMM]}\right), \quad \text{for } t < 7 \text{ days} \tag{6.57}$$

$$\mu_d = \frac{\mu_d^{max}}{1 + \left(\frac{K_{d,AMM}}{[AMM]}\right)^2}, \tag{6.58}$$

with μ_{max} and μ_d^{max} the maximum growth and death rate, respectively; $K_{\mu,AMM}$ the ammonia growth inhibition constant, and $K_{d,AMM}$ the ammonia death inducing constant.

The lysis rate μ_l was empirically introduced as a fitting parameter to provide a mean to remove dead cells from the system in order to correctly describe the behaviour of the cell viability during cultivation.

In a next step, the ammonia concentration can be described as a function of the cellular-growth-rate, and maintenance-dependent parameters (Jang & Barford 2000) as follows:

$$\frac{dV_R\left[AMM\right]}{dt} = q_{AMM} V_R X_V - Q_{out}\left[AMM\right], \tag{6.59}$$

with q_{AMM} to be

$$q_{AMM} = \frac{\mu}{Y_{\mu,AMM}} + m_{AMM}, \tag{6.60}$$

where $Y_{\mu,AMM}$ represents a growth-dependent ammonia yield coefficient, and m_{AMM} represents a maintenance-related coefficient.

In addition, the model includes the mass balance for the target mAb. In the case of Villiger et al. (2016b), the specific productivity q_{mAb} of the employed cell line was observed to be constant with respect to the integral viable cell densities, but it followed a bell-sharped curve as a function of the extracellular pH

(Ozturk & Palsson 1991). The specific productivity can be consequently described with a Gaussian distribution with optimal culture pH at about 7.15, as experimentally determined:

$$\frac{dV_R c_{mAb}}{dt} = q_{mAb} V_R X_V - Q_{out} c_{mAb} \tag{6.61}$$

$$q_{mAb} = q_{mAb}^{max} exp \left[-\frac{1}{2} \left(\frac{pH - pH_{opt}}{\omega_{q,mAb}} \right)^2 \right], \tag{6.62}$$

where $\omega_{q,mAb}$ describes the width of the specific productivity as a function of pH.

Moreover, the cell culture model has to include the mass balances of the nucleotide sugars. There are two main classes of nucleotide sugars. A first group that is typically produced in the cytosol – such as uridine diphosphate N-acetylglucosamine (UDP-GlcNAc), guanosine diphosphate fucose (GDP-Fuc) and uridine diphosphate galactose (UDP-Gal) – and the second that is synthesised in the cell nucleus that includes cytidine diphosphate N-acethyneuraminic acid (CMP-Neu5Ac). After synthesis, the nucleotide sugar precursors are transferred to the ER and to the Golgi apparatus via so-called active anti-transporters. The mass balances for the intracellular metabolites can be expressed in a simplified form with only one growth-dependent and one maintenance-dependent parameter. The general mass balance for nucleotide sugars can be stated as follows:

$$\frac{d \left[NS_k^{cyt} \right]}{dt} = \frac{\mu}{Y_{NS,k}} + m_{NS,k}, \tag{6.63}$$

with a growth-dependent yield coefficient $Y_{NS,k}$ and a maintenance-dependent parameter $m_{NS,k}$. In the case, where a precursor is added within the medium feed, such as galactose, the incorporation rate of that nucleotide sugar, $v_{NS,k}$ is assumed to be proportional to the concentration in the cytosol and the external concentration of the monosaccharide in the medium, referred to as MS_k. Consequently, the extended mass bass balance can be expressed as follows:

$$\frac{d \left[NS_k^{cyt} \right]}{dt} = \frac{\mu}{Y_{NS,k}} + m_{NS,k} + k_{NS,k} \left(\frac{[MS_k]}{K_{NS,k}^{MS}} - \left[NS_k^{cyt} \right] \right), \tag{6.64}$$

where $K_{NS,k}^{MS}$ represents the equilibrium constant describing the equilibrium between the monosaccharide concentration in the medium and in the cytosol.

After collecting data from various micro-bioreactor system runs (including feeding time points, online pH, DO, and sparging rates; analytical results of N-linked glycosylation; and intracellular components), Villiger et al. (2016b) used cell- and process-specific values as initial and boundary conditions for the structured glycosylation model. In addition, the set of unstructured ordinary differential equations of the cell culture model was solved and used as a link for the initiation of the glycosylation model, which resulted in a system of partial differential equations (PDEs). In total, the experimental data of 12 experiments with a total of 48 different glycan structures were used to tune the parameter space of the glycosylation model. Villiger et al. (2016b) tested the predictive capability of the model framework for different feeding strategies, resulting in a vast modulation of the glycosylation patterns, and the model results were able to predict the main glycan structures in both experiments, as shown in the top part of Figure 6.6. Comparing the experimentally measured glycan forms and the model predictions revealed that the described

Model predictions

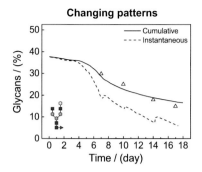

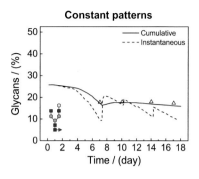

Predictive Capability

Figure 6.6 Evaluation of model performance. Predicted time evolution of an exemplary glycan structure (G1) for two different culture conditions (upper left and upper right), resulting in either a changing or constant glycan pattern as a function of time. The dotted line represents the model prediction for the instantaneous produced glycans, while the solid line represents the cumulative produced glycans. Below, the overall predicted versus measured mAb Fc glycosylation is presented for the most abundant glycan species of the overall data set. © Reprinted from *Biotechnology Progress*, Thomas K. Villiger, Ernesto Scibona, Matthieu Stettler, Hervé Broly, Massimo Morbidelli and Miroslav Soos, 'Controlling the time evolution of mAb N-linked glycosylation: Part II: Model-based predictions', 2016, with permission from John Wiley and Sons

modelling framework was able to predict glycosylation processes for a given process platform (bottom part of Figure 6.6). For fed-batch systems, the N-linked glycosylation profiles of the antibodies change with respect to time due to the accumulation of waste metabolites like ammonia (top left part of Figure 6.6). The model was able to define a feeding strategy that would result in constant glycosylation patterns in time through the addition of galactose and manganese at the latest stages of the culture (top right part of Figure 6.6), indicating that mechanistic models can be used as a robust tool for the optimisation towards defined glycosylation patterns for a given fed-batch process (Figure 6.6). A drawback of this deterministic unstructured cell culture model, which represents the basis of the mechanistic glycan model, is the very limited space of its predictive capability and can be only considered as valid for a stable platform process.

Perfusion Culture for IgG Production Karst et al. (2017b) expanded the application of the dynamic mechanistic N-linked glycosylation model to perfusion cultures. Therefore, a DoE with varying cell density (20×10^6 cells/mL ($-$), 40×10^6 cells/mL (0), and 60×10^6 cells/mL (+)), and supplementation of manganese ($0.01\ \mu M$ ($-$), 0.1 μM (0), and $1.0\ \mu M$ (+)), as well as galactose (0 mM ($-$), 5 mM (0), and 10 mM (+)) was constructed. Increasing the viable cell density represents a valuable tool to vary the ammonia concentration within the culture broth, which has been shown to be one of the most effective inhibitors of glycosylation, while manganese and galactose have been reported to support the production of more complex glycans, such as the galactosylation and sialylation of the terminal glycan units. While Mn^{2+} represents an essential primary co-factor for GnTI, GnTII and GalT, galactose is the sugar precursor of the UDP-Gal nucleotide sugar donor (Gramer et al. 2011, Ramakrishnan et al. 2004). Both high concentration of nucleotide sugar donors and high co-factor concentration are expected to enhance the dynamics within the glycosylation network. Due to the different reactor and process dynamics in perfusion, Karst et al. (2017b) used a modified version of the unstructured cell culture model. The overall reactor balance is the same as in Equation (6.51), but in contrast to the fed-batch reactor, Q_{out} represents the sum of cell-removing bleed and cell-free harvest flowrates

$$Q_{out} = Q_{Harvest} + Q_{Bleed}, \qquad (6.65)$$

which results in the overall mass balance

$$\frac{dV_R}{dt} = Q_{in} - Q_{Harvest} - Q_{Bleed}. \qquad (6.66)$$

Due to continuous media exchange and removal of cells in the bleed stream, the operation of perfusion bioreactors allows to maintain high cell viability close to 100 per cent; therefore, cell death and lysis can be neglected in the overall mass balance of the viable cell density:

$$\frac{d(V_R X_V)}{dt} = \mu V_R X_V - Q_{Bleed} X_V, \qquad (6.67)$$

with μ the growth rate (day^{-1}) and V_R the reactor volume (L), and X_V the viable cell concentration (10^6 cells/mL). At steady state in a perfusion run, the viable cell density is kept constant, and therefore, the volumetric bleed rate can be expressed as follows:

$$Q_{Bleed} = \mu V_R. \qquad (6.68)$$

For the growth rate μ – as well as the mass balances of galactose, manganese, ammonia – and the mAb concentration, Equations (6.52), (6.53), (6.57), (6.59) and (6.61) are still valid but have to be adjusted to the corresponding Q_{out}. With respect to the concentration of the nucleotide sugar donor UDP-Gal, Karst et al. (2017b) used a simplified version of Equation (6.64), which was sufficient to reproduce the experimental data without the addition of more parameters:

$$\frac{d\left[UDP - Gal_{Gal}^{cyt}\right]}{dt} = m_{UDP-Gal,Gal}$$

$$+ k_{UDP-Gal,Gal}\left(\frac{[MS_{Gal}]}{K_{UDP-Gal,Gal}^{Gal}} - \left[UDP - Gal_{Gal}^{cyt}\right]\right). \qquad (6.69)$$

As in the case of the fed-batch model, the parameters derived from the unstructured perfusion model provide the initial and boundary conditions for the structured N-linked glycosylation model. Nucleotide and nucleotide sugar concentrations (except for UDP-Gal) were assumed to be constant during the investigated conditions since continuous systems are known to provide a steady-state condition both intra cellularly and extracellularly. As illustrated in the section about mechanistic N-linked glycosylation model, the kinetics of the enzymatic reactions – in particular, Equations (6.20) and (6.23) were linked to the concentrations of Mn^{2+} and AMM, as well as to the availability of the sugar precursors. This enabled the simulation of the effect of the different operating conditions on the N-linked glycosylation patterns. Model-specific constants and other Golgi-specific parameters were taken from the literature (Jimenez del Val et al. 2011, Villiger et al. 2016b).

With respect to the experimental investigation, a DoE approach was combined with the sequential screening of several set points within long-term perfusion cultures. The perfusion runs were performed at a fixed harvest rate of 1.0 RV/day, and VCD set point changes were enabled by varying the fraction of base to enriched media between 100/0 per cent (20×10^6 cells/mL), 65/35 per cent (40×10^6 cells/mL) and 30/70 per cent (60×10^6 cells/mL), respectively, to account for higher metabolite consumption. The DoE was based on a two-level full factorial design, which was carried out in sequential perfusion culture experiments. In Table 6.1, the experimental conditions of the full factorial design with varying VCD, galactose and manganese concentration as well as the resulting patterns of the main glycoforms (such as HM, G0, G1 and G2) are summarised.

The experimental results revealed a high variability of glycoforms as a function of the operating conditions as shown in Table 6.1. While high-mannose (HM) structures were, in particular, found at higher VCDs due to the higher ammonia concentrations, ungalactosylated and galactosylated structures were widely spread as a result of the different operating conditions. With respect to the DoE conditions, for G0, a modulation space was found between 36 per cent (minimum) and 70 per cent (maximum), while G1 ranged from 18 to 48 per cent, and G2 from 1 per cent to 11 per cent. The systematic modulation approach was then, in a next step, supported by the training of the mechanistic glycosylation model, based on the first eight DoE experiments. The mechanistic model showed good agreement with the experimental measured glycan structures – in particular, for highly modulated structures, such as G0 and G1. The model was capable of reproducing both the expected inhibitory effect of ammonia/high VCD and the enhancing effect of manganese and galactose, as shown in Figure 6.7. Six cell-specific parameters of the unstructured cell culture model were fitted: the maximum cellular growth rate, μ_{max}; the ammonia growth inhibition constant, $K_{\mu,AMM}$; the ammonia and antibody specific productivity, q_{AMM} and q_{mAb}; and the maintenance coefficient of UDP-Gal $m_{UDP-Gal}$; as well as the the equilibrium constant of the galactose UDP-Gal equilibrium $K_{UDP-Gal,Gal}^{Gal}$.

On the other hand, the mechanistic glycosylation model required the estimation of seven parameters to correctly reproduce the glycosylation profiles within the entire DoE: the dissociation constants for Golgi resident enzymes and the term N_A^{Golgi} in Equation (6.50) that represents the transport of ammonia to the Golgi apparatus.

In a next step, the model was used to predict an independent set of experimental conditions referred to as Centre, Outlier and Edge in Table 6.1. The trained model was, indeed, able to predict the resulting glycosylation patterns and, therefore, showed the capability for both intrapolation and extrapolation from the operating

Table 6.1 Experimental conditions of the full-factorial DoE as well as a middle point, an outlier and an edge point of the DoE with varying VCD, Gal and Mn^{2+} concentrations performed in sequential long-term perfusion cultures and the resulting steady-state glycan distributions in terms of the main glycans – such as HM, G0, G1 and G2. © Reprinted from *Biotechnology and Bioengineering*, Daniel J. Karst, Ernesto Scibona, Elisa Serra, Jean-Marc Bielser, Jonathan Souquet, Matthieu Stettler, Hervé Broly, Miroslav Soos, Massimo Morbidelli and Thomas K. Villiger, 'Modulation and modeling of monoclonal antibody N-linked glycosylation in mammalian cell perfusion reactors', 2017, with permission from John Wiley and Sons.

DoE conditions	VCD / (10^6 cells/mL)	Gal / (mmol/L)	Mn^{+2} / (μmol/L)	HM / (%)	G0 / (%)	G1 / (%)	G2 / (%)
− − −	20	0	0.01	1.2	61.1	31.7	3
− + −	20	10	0.01	1.1	49	40.9	5.6
− − +	20	0	1	1.4	40.6	45.4	8.6
− + +	20	10	1	1.6	35.7	47.7	10.9
+ − −	60	0	0.01	6.2	70.6	18.1	1.1
+ + −	60	10	0.01	3.7	64.5	26.3	2.2
+ − +	60	0	1	2.8	47.6	39.4	6.1
+ + +	60	10	1	3.1	43.8	41.2	6.8
Center	40	5	0.1	1.9	38.3	46.2	9
Outlier	40	20	2	1.1	31.3	50.6	12.7
Edge	40	0	0.01	2.8	62.8	28.2	1.1

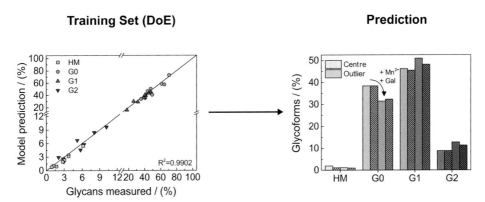

Figure 6.7 Evaluation of model performance in a perfusion application. On the left, the overall predicted versus measured mAb Fc glycosylation for the training set (two-level full-factorial DoE) is shown for the most abundant glycan species of the overall data set of Karst et al. (2017b). The trained model was used in a second step to predict the glycan patterns for another set of operating conditions. On the right, the experimental (plain) and predicted (stripped) glycan distribution at the centre (green) and the outlier point (point) in Table 6.1 are compared.

conditions covered by the DoE. For illustrative purposes, the experimental and predicted glycosylation patterns for the Centre and the Outlier operating conditions are compared in Figure 6.7.

From a more general point of view, these studies confirm the strength of mechanistic models with respect to process development, also in the case of biological systems, and they underline the power of the combination of unstructured and structured models. Glycan processing is a precisely described sequence of enzymatic steps that shows a high sensitivity with respect to environmental conditions. By combining this model with an unstructured cell culture model that is often kept rather simple, it is possible to train a mechanistic model that predicts N-linked glycosylation for both fed-batch and perfusion processes. Of course, the model predictions are only valid for suitable operating ranges of the given fed-batch or perfusion platform process. Nevertheless, this fact is important with respect to perfusion process development, which was discussed in Chapter 4. These examples underline the possibility of fine-tuning product quality of a given platform process with proper feed adjustments.

6.6 STATISTICAL AND HYBRID MODELLING

In this section, we discuss the application of statistical and hybrid modelling approaches to cell culture processes. The latter approach combines mechanistic or knowledge-driven models, as described in the previous section, with statistical or data-driven techniques. While mechanistic modelling approaches are derived from physical, chemical or biological principles, statistical approaches, often referred to as *black-box* or *empirical* models, search for data-driven correlations between input and output variables without involving any consideration based on first principles. More advanced approaches of this type, including time series analysis and historical prediction models, have also been developed to describe process dynamics.

Data-driven approaches have been proven to accurately express process trajectories in many areas of engineering, including bioprocessing (Bailey & Ollis 1986, Mercier et al. 2014). The data typically available in bioengineering and biotechnology feature all of the four Vs of big data characteristics (Narayanan et al. 2019b) – namely volume, due to high-throughput and spectroscopic data; velocity, due to online measurements; veracity, as a result of the significant noise due to the living process nature and to the limited control, particularly at the small experimental scale; and variety, as a result of the heterogeneity of processes and products as well as devices, apparatuses and data management systems used along the lifecycle of process development. Moreover, the relationships between biological variables often are not completely understood, are not thoroughly measured and do not follow simple rate laws. Therefore, statistical methods provide an attractive way to efficiently extract information from such complex data and are a valuable tool to provide first insights into the bioprocess, highlight important relationships and influential variables, and generate hypotheses for subsequent modelling approaches (Sokolov et al. 2018). Statistical modelling offers a versatile toolbox ranging from simple multiple regression for uncorrelated regressors, multivariate analysis accounting for correlation amongst the regressors and non-linear machine learning approaches.

Despite the limited structural relation to the biotechnological process, the successful generation of these models depends on a clear understanding of the underlying model definitions and assumptions as well as the possibilities to tailor these to the nature of bioprocess data.

Statistical models are often used to support experimental design as well as to analyse the corresponding experimental data. The so-called design of experiments (DoE) approach facilitates the identification of relationships between given input and output variables, while placing experimental conditions as suitable combinations of several input variables at different levels. Such a DoE approach is significantly more efficient with regards to the experimental effort compared to classical approaches testing one variable at a time (Sommeregger et al. 2017). Moreover, combined variable testing is more likely to find important interaction or limitation effects amongst tested variables. This approach can also be utilised in a sequential way to iteratively identify relevant process variables and eventually define the so-called process design space in which these shall be controlled to maintain the targeted product quality in a defined specification range (Politis et al. 2017, Steeno 2010, Von Stosch & Willis 2017). In general, such quantitative approaches supporting the experimental effort during process development as well as providing new routes for knowledge generation are highly supported by the PAT and QbD initiatives of the regulatory authorities (Mercier et al. 2016, Rathore et al. 2016). However, solely statistically driven DoEs can be extremely laborious due to the amount of information which has to be tested (input variables and their ranges) and the limitations of the underlying linear models to withdraw relevant information on non-linear dynamic processes. Therefore, prior knowledge on central variables, their ranges and their interrelationships as well as advanced techniques to account for the non-linear and multivariate process nature are highly relevant to efficiently use experimental resources and already-available information. Hence, several applications and improvements of the classical DoE approaches have been presented in the literature in order to decrease the experimental effort and build more reliable and more generically applicable model solutions (Brühlmann et al. 2017b, Chakrabarty et al. 2013, Von Stosch & Willis 2017).

Besides the classical *white-box* (mechanistic) and *black-box* (statistic) modelling approaches, so-called *grey-box* or *hybrid* models present a compromise between mechanistic and statistic strategies by combining the available knowledge (such as simple unstructured cell culture models) with data-oriented modelling techniques and have been shown to often provide better predictions than pure black-box or pure white-box approaches (Narayanan et al. 2019a).

6.6.1 Multivariate Data Analysis Tools

In the case of large data sets with various independent variables, multivariate data analysis (MVDA) represents a useful tool to handle the high dimensionality and correlation between the different variables. Multivariate data analysis basically includes a set of tools that deal with more than one independent variable at a time and aim to perceive the correlations between such variables. In particular, principal component analysis and partial least square regression analysis extract important information, simplify datasets and estimate relationships between process variables (Kettaneh et al. 2005, Mercier et al. 2014). These techniques have been applied in investigations such as peptone screening batch fault detection, cell culture media screening and scale-down model qualification (Brunner et al. 2018, Gunther et al. 2007, Luo & Chen 2007, Ryan et al. 2010, Sokolov et al. 2017c, Tsang et al. 2014), and are discussed in the next sections.

Principal Component Analysis Extensive data sets, including the measurements of many variables, are increasingly common in cell culture processes due to the availability of more advanced process monitoring tools including omics approaches, high-throughput systems and online monitoring. Principal component analysis (PCA) is a common tool to reduce the number of variables into a smaller set of orthogonal principal components (PCs) while retaining most of the information of the original data set and separating it from process noise. PCA generally can be used as a visualisation tool to help identify groups, trends and central interrelationships (Jolliffe 2005, Meglen 1992). We can summarise the four goals of PCA as follows:

1. Extraction of the most important information from the data
2. Simplification of the visualisation and description of the data set benefiting from the highly correlated process nature
3. Analysis of the underlying correlation structure of the observations and the variables
4. Compression of the data dimensions and noise level by keeping only the important information

In particular, PCA decomposes the data matrix X with M variables and N observations. The data set can be represented either as M N-dimensional vectors $x_1, \ldots, x_M$ or in a $(N \times M)$ matrix, X.

The goal of PCA is to reduce the dimension from M to A, with A typically much smaller than M, while retaining maximal information content. This can be done by finding linear combinations of the original variable x_m ($m = 1, 2, \ldots, M$) with the coefficients $p_{m,a}$ ($a = 1, 2, \ldots, A$), called principal components (PCs). These PCs are orthogonal to one another and are derived such that they explain maximal variance of the given data sets. By solving this maximisation problem, it can be found that p_a ($a = 1, 2, \ldots, A$) are the eigenvectors of the covariance matrix, S, of the data corresponding to the A largest eigenvalues λ (Jolliffe 2005). The elements of the eigenvectors p_a are commonly called *loadings*, and the elements of the linear combinations Xp_a ($= t_a$) are called the *scores*. From a graphical perspective, an eigenvector represents the identified direction along which the maximal fraction of the variance, not explained by the previously defined eigenvectors, can be explained, which is indicated by the corresponding eigenvalue. In particular, the first and the second principal component represent the eigenvectors having the highest eigenvalues, and consequently contain the most information – e.g., the axis along which the data is most spread out.

For PCA, the components can be obtained by performing a single-value decomposition (SVD) of the data matrix X as follows (Bartholomew 2010):

$$X = U\Delta V^T + E, \tag{6.70}$$

with U representing the $(N \times A)$ matrix of the left singular vectors and V representing the $(M \times A)$ matrix of the right singular vectors, often referred to as loading matrix (columns of V represent the principal directions) and Δ representing the diagonal matrix of the non-zero singular values with square roots of the $X^T X$ matrix as elements and the residual E.

A common representation in chemometrics is the decomposition of the matrix X as the sum of the outer product of vectors score t_a and loadings $p_a a$ (Kumar et al. 2014).

$$X = t_1 p_1^T + t_2 p_2^T + \cdots + t_A p_A^T + E \quad or \quad X = TP^T + E. \tag{6.71}$$

Here, the single difference from Equation (6.70) is the fact that V is identical to P but scaled to length one. A convenient way to generate the matrices T and P is based on non-linear iterative partial least square (NIPALS) applied to X, which is scaled to unit variance along each column. As described by Kumar et al. (2014), p_a^T can be defined as follows, with the score vector t_a selected from the column of matrix X that has the largest variance:

$$p_a^T = \frac{t_a^T X}{t_a^T t_a}. \tag{6.72}$$

Then, p_a^T can be normalised to unit length when multiplying by a constant factor c:

$$c = \frac{1}{\sqrt{p_a^T p_a}}. \tag{6.73}$$

The following equation is then used to find the new score vector:

$$t_a = \frac{X p_a}{p_a^T p_a}. \tag{6.74}$$

The new scores are then used to compute the loadings using Equation (6.72) and the procedure (Equations (6.72–6.74)) is repeated until convergence for the ath principal component. Subsequently, in order to compute the next principal component, the X matrix is deflated using $X = X - t_a p_a^T$. Equations (6.72–6.74) are applied using the deflated matrices to compute the subsequent principal components. Therefore, the residual E is iteratively decreased by adding new principal components along directions explaining largest variance.

Following one of the central goals of PCA, namely to simplify the visualisation and interpretation, in most of the cases, the high-dimensional data are then visualised in 2D plots of the lower-dimensioned scores to identify characteristic groups or of the lower-dimensioned loadings to study the characteristic interrelationship among the

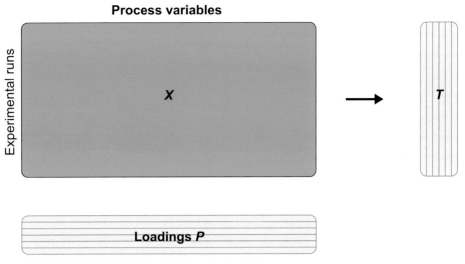

Figure 6.8 Simplified PCA on process data matrix X. The relation of the new system to the original one is given by the scores T (projection of the experiments) and the loadings P (directions of the original variables in the new space).

Figure 6.9 Exemplary score plot for the first two principal components for three process conditions (1, 2 and 3), the axes indicate the variance explained by the corresponding principal component. In this exemplary case, the resulting clusters allow the clear distinction of different process conditions in the experimental data set.

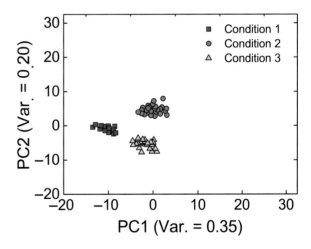

variables. An example of such a score plot is presented in Figure 6.9, where the first two PCs which jointly explain 55 per cent of the variability in the data are presented.

This data set was obtained in an analysis of the intracellular metabolite profile during sequential perfusion cultures. For this purpose, Karst et al. (2017c) measured highly dimensioned MALDI-TOF MS spectra of intracellular metabolites for the sequence of three VCD set points (20×10^6, 60×10^6 and 40×10^6 cells/mL) over 10 selected culture days. Applying PCA to these spectra revealed, as shown in Figure 6.9, a clear clustering of the three set points in the score plots of the first three principal components. Such cluster specific grouping can be used as an indication of a set-point specific intracellular metabolite profile and shows the visualisation potential of PCA. It can also be very supportive in identifying abnormal behaviour along the dynamic evolution of an experiment (Sokolov et al. 2015).

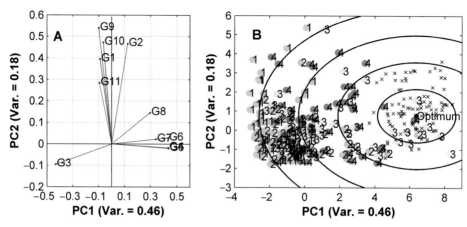

Figure 6.10 *Y*-PCA of glycoforms. (A) Loading plot of the 11 *Y*-variables; (B) the score plot is labelled according to the utilised clone and additionally shows the location of the optimum as well as of 100 projected corners of the specification range (cross markers). The ellipses visualise equidistant Mahalanobis distances from the optimum ranging from 1 to 4 U. © Reprinted from *Biotechnology Progress*, Michael Sokolov, Jonathan Ritscher, Nicola MacKinnon, Jean-Marc Bielser, David Brühlmann, Dominik Rothenhäusler, Gian Thanei, Miroslav Soos, Matthieu Stettler, Jonathan Souquet, Hervé Broly, Massimo Morbidelli and Alessandro Butté, 'Robust factor selection in early cell culture process development for the production of a biosimilar monoclonal antibody', 2016, with permission from John Wiley and Sons

The possibility to jointly interpret scores and loadings, which – given the mathematical definition – follow the same orientation, shall be demonstrated on a cell culture process development study for a biosimilar. Here, Sokolov et al. (2017b) applied PCA to 11 (scaled) final product quality variables shown in the loading plot in Figure 6.10A, where one can distinguish three groups of variables. These can be identified as three groups of lower, intermediate and higher processed glycoforms along the glycosylation network (see also Section 6.5.2). The score plot demonstrates the location of the final product quality of the performed experiments under different conditions with respect to the target product (highlighted by optimum and cross-marks for the specification range). Given the joint orientation, one can identify in a single experiment that to reach optimal biosimilarity, the quantity of the forms G5 to G8 has to be increased (large values along PC1), while G3 and the other glycoforms have to decreased (small values along PC2).

Partial Least Square Regression Partial least squares (PLS) analysis represents another important multivariate technique used for multivariate regression of one (PLS1) or several (PLS2) dependent variables (e.g., final product quality) based on correlated independent variables (e.g., process variables). The multivariate nature of the model enables the reduction of the dimensions of the interrelationship between, e.g., the process conditions and the product characteristics, to simplify interpretation. Hence, unlike PCA, PLS is not a visualisation but a regression tool enabling the prediction of one or several output variables based on several input variables. The major preparation steps for PLS are definition of input/independent variables x, and output/dependent variables y, problem definition, data preparation including alignment and unfolding (as product quality measurements usually less frequent then process measurements), as well as data pretreatment to ensure comprehensible variable comparison (Sokolov et al. 2015). The set of X in bioprocessing can, e.g., comprise measurement data of several process variables (such as VCD, GLC, LAC, pH, DO) or spectral data, while Y contains the targeted response variables, such as final titre concentration or product quality patterns, or reference measurements when spectral data are considered.

The X and Y data matrices are mean centred and scaled before applying the PLS regression. PLS essentially finds a set of orthogonal latent variables that best predicts Y as well as best models the X matrix. The number of latent variables (LVs) are typically much smaller than the number of independent variables (X). PLS aims at finding two sets of coefficients or weights namely W^* and C to create a linear combination of the X and Y columns, respectively, such that these linear combination have a maximum covariance. These linear combinations are called the X-scores (denoted by T) and the Y-scores (denoted by U) computed as shown here:

$$T = XW^* \tag{6.75}$$

$$U = YC. \tag{6.76}$$

Inner relations of the PLS regression model link the X and Y matrices together while the outer relations shown here describe them individually:

$$X = T \times P^T + E \tag{6.77}$$

$$Y = U \times C^T + F, \tag{6.78}$$

with P^T being the loading matrix of the X space, C^T the loading matrix of the Y space, and E and F the residual matrices of the X and Y spaces, respectively. X-scores, or T, are also good predictors for Y variables, as seen in the following equation:

$$Y = T \times C^T + G, \tag{6.79}$$

Using Equations (6.75) and (6.79), this then becomes

$$Y = XW^*C^T + G = XB + G, \tag{6.80}$$

where

$$B = W^*C^T, \tag{6.81}$$

with B being the PLS regression coefficient and G the residual matrix. Similar to the PCA, a NIPALS algorithm is used to compute the T, U, C, P and W^*, as nicely illustrated by Wold et al. (2001).

The portions of the model that cannot be explained are called residuals, and their value is useful to understand the applicability of the model. A good model can be distinguished from a poor model because it will feature low residual values. Residual matrices can be used to recalculate PLS components after the first model, and this can be done until 99 per cent of the variance can be explained. The number of significant PLS components required for a calibration model can be estimated using cross-validation (Cao et al. 2010, García-Muñoz & Polizzi 2012, Indahl 2005).

When training a PLS model, it is important to avoid overfitting and include only as many latent variables as are required to describe general patterns affecting the entire data set and not only very small subsections of it, which would limit later

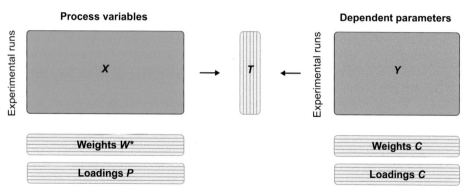

Figure 6.11 Simplified PLS from, e.g., process data matrix X to Y matrix containing product quality information. The high-dimensional relationship is reduced to a lower-dimensional number of latent variables, consisting of the scores T (projections of the experiments) and the corresponding loadings W^* and C (transformations of the original X and Y variables to the latent variable space).

its use to predicting new data. For this purpose, a technique called cross-validation is used, which iteratively trains models on subsets of the data and determines the optimal numbers of latent variables (Cao et al. 2010).

As a measure of model performance during calibration, the root mean square error in cross validation (REMSECV) can be defined as follows:

$$\text{RMSECV} = \sqrt{\frac{1}{n} \sum_{i=1}^{n} \left(y_{test,i} - \hat{y}_{text,i}\right)^2}. \tag{6.82}$$

with n representing the number of observations in the calibration set, which is in groups of K used as an internal test set so that each observation i is tested as $y_{test,i}$ and estimated as $\hat{y}_{test,i}$ by the corresponding model on the subset used for training. Such RMSECV is calculated for each additional LV, which results in an optimal number finding a trade-off between explaining the complex interrelationship between X and Y while not overfitting it. In addition, the relative variance explained in cross-validation (Q^2) can be considered, which is described as the ratio of explained (ESS) to total variance (TSS) in the internally predicted data.

$$Q^2 = 1 - \frac{\text{ESS}}{\text{TSS}}. \tag{6.83}$$

$Q^2 = 1$ signifies a perfect prediction model. This characteristic is maximal at the optimal number of LVs used. For a proper evaluation of the external prediction power, however, the best calibrated model has to be evaluated with an external data set, which was not part of the cross-validation procedure and is therefore completely unknown to the model. This step allows the testing of the robustness and extrapolation potential of the model towards new data. The root mean square error in prediction (RMSEP) is one of the most common performance indicators for this:

$$\text{RMSEP} = \sqrt{\frac{1}{n} \sum_{i=1}^{n} \left(y_{ext,i} - \hat{y}_{ext,i}\right)^2}, \tag{6.84}$$

with n representing the number of observations in the corresponding external set, $y_{test,i}$ representing the y-value of the ith observation in the external set and $\hat{y}_{ext,i}$ representing the model estimation of the y-value of the ith observation.

PLS regression analysis represents a widely used technique in bioprocess data analysis. It has been used, for example, to predict and forecast final product quantity and quality attributes – such as titre, aggregates, fragments, charge variants and N-linked glycosylation in high-throughput cell culture experiments, across multiple other reactor scales (Sokolov et al. 2017c, 2018). Moreover, it is a widely used technique to build predictive models for relevant process variables (Y) based on spectra acquired, e.g., Raman or NIR sensors (Feidl et al. 2019).

As a detailed example, Sokolov et al. (2017c) investigated the results of 91 experimental fed-batch runs, all performed in the ambr15® system. The runs were based on four experimental blocks of DoEs with the objective to test nine different media supplements and three different process-execution strategies. The data were separated in three different characteristics blocks: Block Z, the experimental conditions, was spanned by the experimental runs (rows) and different process conditions (columns); Block X, the process dynamics, was built by the experimental runs (rows) measured process variables at different time points (columns) and Block Y represents the different final product quality attributes (columns) for the experimental runs (rows). Such separation exists not only to enable the learning of the characteristic interrelationship of the process to the product but also enables the identification of the Z parameters driving the process (which are important for process control) and X variables best representing the dynamic process behaviour (which are important for process monitoring). In the first step, a principal component analysis was used and revealed the strong correlation characteristics among the product quality attributes, similarly to Figure 6.10A. In the second step, partial least square regression was applied to predict product quality attributes, revealing the interrelationship of process characteristics and product quality variables. It shall be highlighted that here, given the strong identified correlation structure in Y, PLS2 models were used, linking the process to the product behaviour with a single model (instead of one model per product variable). This remarkably simplified the interpretation of the analysis results. Thereby, several models were built with different portions of process history incorporated – i.e., forecasting from the initial and designed conditions Z – as well as models incorporating some or all of the dynamic process history X.

Figure 6.12 presents the Q^2 obtained for such predictions for the different variables (columns) and different amounts of historic information used (rows). First of all, it shall be highlighted that the models can predict most of the quality variables very decently with Q^2 above 0.7. Even further, one can observe that LMW could be decently predicted based on using the initial conditions Z, while for variables such as the charge variant C4, at least information until culture day (T5) was required. This can be explained by the fact that several product quality attributes are allowing the prediction of these attributes early on in the process, while other attributes were more affected by process parameters, such as pH and temperature shifts, that were applied during the culture. This study highlights the strength of statistical models for large correlated data sets, to identify important dynamic patterns and process drivers to support decision making in process development.

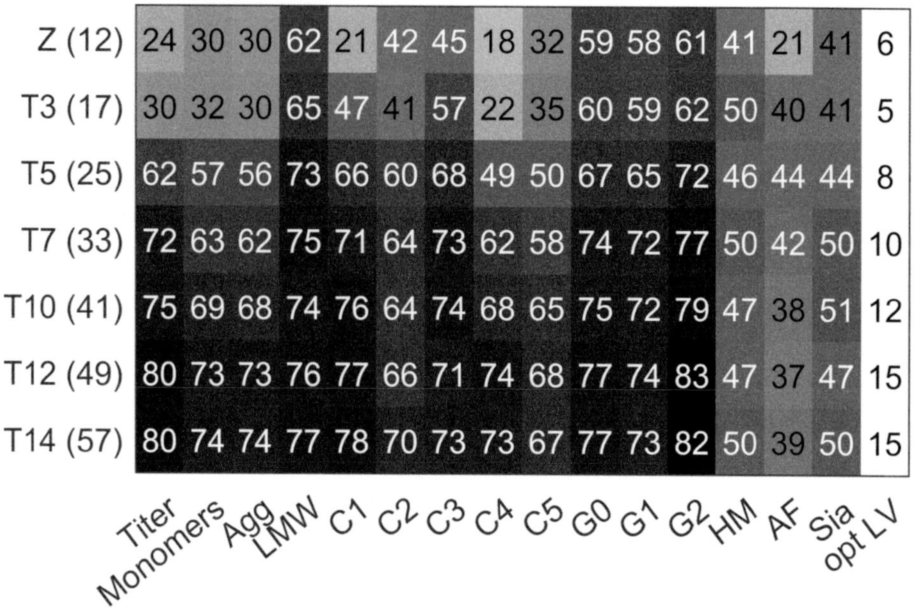

	Titer	Monomers	Agg	LMW	C1	C2	C3	C4	C5	G0	G1	G2	HM	AF	Sia	opt LV
Z (12)	24	30	30	62	21	42	45	18	32	59	58	61	41	21	41	6
T3 (17)	30	32	30	65	47	41	57	22	35	60	59	62	50	40	41	5
T5 (25)	62	57	56	73	66	60	68	49	50	67	65	72	46	44	44	8
T7 (33)	72	63	62	75	71	64	73	62	58	74	72	77	50	42	50	10
T10 (41)	75	69	68	74	76	64	74	68	65	75	72	79	47	38	51	12
T12 (49)	80	73	73	76	77	66	71	74	68	77	74	83	47	37	47	15
T14 (57)	80	74	74	77	78	70	73	73	67	77	73	82	50	39	50	15

Figure 6.12 Variance explained in cross-validation (Q2 in %) for PLS2 models for different CQAs at day 14 (columns) based on different lengths of process history utilised (rows, where the first one represents the controlled process conditions and the further ones refer to incorporating the complete process history). The numbers in brackets indicate the number of X variables used in the corresponding PLS2 model. © Reprinted from *Biotechnology Progress*, Michael Sokolov, Jonathan Ritscher, Nicola MacKinnon, Jonathan Souquet, Hervé Broly, Massimo Morbidelli and Alessandro Butté, 'Enhanced process understanding and multivariate prediction of the relationship between cell culture process and monoclonal antibody quality', 2017, with permission from John Wiley and Sons

6.7 HYBRID MODELLING APPROACH

Having shown the strengths of both modelling approaches, we have seen that while deterministic models require a sufficient understanding of the underlying mechanisms, statistical models require sufficiently large data sets to understand all the correlations for a given set of data. Hybrid modelling approaches represent a compromise of these two techniques by combining in a pragmatic way *a priori* knowledge of deterministic models as a structural backbone with the flexibility of a statistical model to describe unknown interrelationships inside that structure (Chen et al. 2004, Glassey & von Stosch 2018, Narayanan et al. 2019c, Von Stosch et al. 2016). A very common approach in hybrid modelling of bioprocesses is to apply statistical tools to estimate rate constants of a given deterministic reaction network or cell culture model, such as the cellular growth constant or specific metabolite production and consumption rates using either artificial neural networks or PLS models (Teixeira et al. 2005, Von Stosch et al. 2016).

An important example of hybrid modelling in bioprocessing was presented by Narayanan et al. (2019c). In this work, the added value of hybrid modelling compared to the alternative (particular statistical) approaches was presented from several perspectives. The analysis was performed on a cell culture process characterisation data set with almost 80 runs (Sokolov et al. 2015). As presented in Figure 6.13A, the hybrid modelling technique could significantly outperform statistical modelling

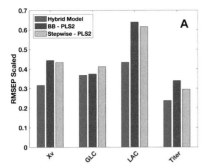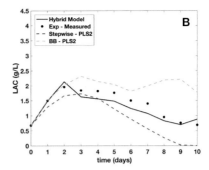

Figure 6.13 (A) Comparison of the accuracy of different models reported in terms of RMSEP scaled with respect to the standard deviation for X_v, glucose, lactate and titre and (B) shows the evolution of lactate for one run predicted by different models. © Reprinted from *Biotechnology and Bioengineering*, Harini Narayanan, Micheal Sokolov, Massimo Morbidelli and Alessandro Butté, 'A new generation of predictive models: The added value of hybrid models for manufacturing processes of therapeutic proteins', 2019, with permission from John Wiley and Sons

techniques to predict the complete evolution profiles of central variables such as lactate and titre. It shall be highlighted here that the goal of hybrid modelling in this analysis was to reliably predict the process behaviour on the designed variables only (Z variables). This shall enable one to define the process design space and reliable digital process twins to simulate process behaviour. The lack of available dynamic measurements is compensated by mechanistic process knowledge, while, for the statistical models, a day-wise propagation procedure is applied. Figure 6.13B demonstrates such enhanced predictive power for lactate. This predictive accuracy is essential to reduce the experimental effort and support experimental design in process development based on advanced models reflecting prior knowledge. Besides the potential to reduce the experimental design and more efficiently converge to an optimal design space, this study has additionally highlighted the important underlying capability for extrapolation clearly going beyond the limits of purely data-driven approaches.

6.8 ADVANCED PROCESS MONITORING AND CONTROL

After discussing the different types and applications of mathematical models, in this section, we elaborate on the use and application of these modelling techniques for process monitoring and control. The PAT/QbD initiative started in 2004 has resulted in worldwide increasing activity in the development of analytical and control strategies in bioprocesses (Glassey et al. 2011; Rathore 2014; Read et al. 2010a, 2010b). The recent advances in process automation and digitalisation technologies increasingly permeate the field of biotechnology, making experimental data from bioreactor runs more accessible and simpler to treat (Carrondo et al. 2012). The implementation of online spectral sensors, such as Raman or NIR, measuring thousands of wavelengths every few minutes, also produce significant amounts of additional data, thus providing new tools for advanced process monitoring and control.

We have seen in the previous section that multivariate data analysis methods are perfectly suited for data processing, reducing data dimensions and noise disturbance, identifying important patterns and trends as well as constructing predictive models. Thereby, statistical regression models or even models based on advanced mechanistic modelling could be used for developing soft sensors based on online Raman or NIR spectra (Feidl et al. 2019, Sommeregger et al. 2017). The spectra from biological samples are, in fact, complex and diverse, and interpreting them to extract the desired information is difficult. Therefore, after proper data preprocessing – which is not discussed in detail here, although it is an important step to be executed with great care – PCA and PLSR analysis can be used to build predictive models that can potentially enable the monitoring of various process variables ranging from measurable species – such as VCD, GLC or mAb concentration – to non-measurable species, such as critical product quality attributes (Escandar et al. 2006, Gautam et al. 2015, Hubert & Engelen 2004). In addition, the application of these tools combined with tools enabling process modelling and optimisation, such as the hybrid model presented earlier, can enable the proactive identification of risks of process failure much earlier – e.g., when certain process parameters start to move out of the process design space – and thus allow the operator to take proper actions.

It appears that such model-based techniques can cover a versatile set of supportive tasks ranging from descriptive (dimension reduction and trend monitoring), diagnostic (alert systems and failure analysis), predictive (soft sensors, forecasting models) and prescriptive (process control and optimisation) applications (Narayanan et al. 2019a). With regards to the ultimate goal in biopharmaceutical manufacturing, namely that of consistently providing product quality within specifications, these tools can be further improved so as to implement a direct control of the critical quality attributes. Such implementation of predictive quality coupled with process control shall eventually advance towards adaptive solutions, shaping the path towards real-time release or at least release by acceptation. Besides data analytics and process modelling tools, data management and integration are crucial to realising the potential of process digitalisation and automation (Luttmann et al. 2015).

The current design of manufacture facilities uses either a *modular approach*, where each unit operation is considered as a separate independent unit that is optimised independently and then integrated with another module without any modification, or an *adaptation approach*, where one- or two-unit operations to be integrated are adapted in a manner that allows their seamless integration. This situation is fundamentally changed in integrated continuous biomanufacturing, where the entire manufacturing facility is basically optimised as a single large unit, thus taking advantage of all the interconnections amongst the single units to design their individual operations to optimise the behaviour of the entire manufacturing facility. This implies overcoming the traditional separation between upstream and downstream units and the collection and elaboration of the data coming from all sensors on different units in a single place, as illustrated in the schematic of a continuous integrated unit shown in Figure 6.14 (Konstantinov & Cooney 2015, Steinebach et al. 2017). For example, the performance of the capture step in a mAb production facility is significantly affected by the behaviour of the perfusion bioreactor. Accordingly, the integration of these two units can be most beneficial (Karst et al. 2017a). For this, it is necessary to monitor the output of the perfusion bioreactor and use it to control the capture step. The bioreactor is, in fact, more prone to deviations in its behaviour as a function of time due to inherent complexities of the culture process, such as cell ageing or metabolic changes leading to decreasing titre values. In such cases, real-time monitoring of product concentration and quality is necessary to adapt the operating conditions of the capture unit in order to keep the desired overall performance. This has to be done at two levels: the highest one, where the product has to be kept within specifications (otherwise, it will have to be discarded), and a low one, where the cost

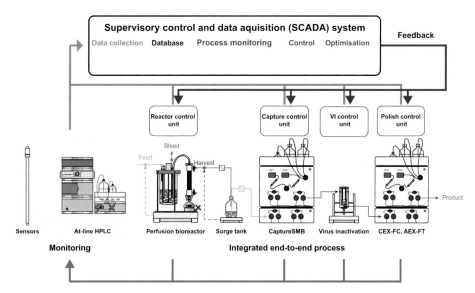

Figure 6.14 Schematic representation of an integrated continuous process with the implementation of sensors and at-line HPLCs for process monitoring. The local control loops of each unit (bioreactor, capture, polish, analytics, spectral data) are connected to a supervisory control and data acquisition system that collects all the data and applies proper data processing steps, in order to perform process monitoring, control and optimisation

of the operation has to be minimised, such as reducing the amount of buffer used per unit of mAb produced. An example of this has been provided by Karst et al. (2017a) where an at-line HPLCs (ProtA, SEC) was used to monitor concentration of the target protein and of its aggregates. This same concept can, in principle, be realised at higher level using the online spectroscopic sensors described earlier (Feidl et al. 2019). The next step is to integrate the available technologies into the entire continuous operation train and to build up control and monitoring solutions for the entire process, as illustrated in Figure 6.14 (Feidl et al. 2020).

6.9 CONCLUSION

In this chapter, we have shown the diversity of different quantitative modelling approaches that can be applied to cell culture processes. These cover general models describing stirred tank bioreactors, fed-batch or perfusion, including unstructured models for describing the cell culture growth and metabolism. These can be coupled with more specific models to obtain tools for controlling relevant and difficult details of the process, like the final product quality attributes. An example of this is the introduction of a dynamic mechanistic model of N-linked glycosylation to control and modulate the mAb glycosylation patterns. Moreover, we discussed the use of statistical and hybrid models enabling the analysis of large data sets and the implementation of soft sensors. With respect to the current trends, model-based advanced process monitoring, optimisation, control and automation techniques will represent key enablers of continuous biomanufacturing, as they enable the reduction of the experimental effort and facilitate the integration, operation and understanding of the different and complex process units.

References

Abu-Absi, N. R., Kenty, B. M., Cuellar, M. E., Borys, M. C., Sakhamuri, S., Strachan, D. J., Hausladen, M. C. & Li, Z. J. (2011), 'Real time monitoring of multiple parameters in mammalian cell culture bioreactors using an in-line Raman spectroscopy probe', *Biotechnology and Bioengineering* **108**(5), 1215–1221.

Ahmed, S. U., Ranganathan, P., Pandey, A. & Sivaraman, S. (2010), 'Computational fluid dynamics modeling of gas dispersion in multi impeller bioreactor', *Journal of Bioscience and Bioengineering* **109**(6), 588–597.

Al-Rubeai, M. (2015), *Animal Cell Culture*, Vol. 9, 9th edn, Spinger.

Allison, G., Cain, Y. T., Cooney, C., Garcia, T., Bizjak, T. G., Holte, O., Jagota, N., Komas, B., Korakianiti, E., Kourti, D., Madurawe, R., Morefield, E., Montgomery, F., Nasr, M., Randolph, W., Robert, J. L., Rudd, D. & Zezza, D. (2015), 'Regulatory and quality considerations for continuous manufacturing May 20–21, 2014 continuous manufacturing symposium', *Journal of Pharmaceutical Sciences* **104**(3), 803–812.

Amanullah, A., McFarlane, C. M., Emery, A. N. & Nienow, A. W. (2001), 'Scale-down model to simulate spatial pH variations in large-scale bioreactors', *Biotechnology and Bioengineering* **73**(5), 390–399.

Ansorge, S., Esteban, G. & Schmid, G. (2010), 'On-line monitoring of responses to nutrient feed additions by multi-frequency permittivity measurements in fed-batch cultivations of CHO cells', *Cytotechnology* **62**(2), 121–132.

Antonia, S. J., Larkin, J. & Ascierto, P. A. (2014), 'Immuno-oncology combinations: A review of clinical experience and future prospects', *Clinical Cancer Research* **20**(24), 6258–6268.

Arnold, L., Lee, K., Rucker-Pezzini, J. & Lee, J. H. (2018), 'Implementation of fully integrated continuous antibody processing: Effects on productivity and COGm', *Biotechnology Journal* **14**(2), 1–10.

Arnold, S. A., Crowley, J., Woods, N., Harvey, L. M. & McNeil, B. (2003), 'In-situ near infrared spectroscopy to monitor key analytes in mammalian cell cultivation', *Biotechnology and Bioengineering* **84**(1), 13–19.

Assirelli, M., Bujalski, W., Eaglesham, A. & Nienow, A. (2005), 'Intensifying micromixing in a semi-batch reactor using a Rushton turbine', *Chemical Engineering Science* **60**(8–9), 2333–2339.

Atkins, P., De Paula, J. & Friedmand, R. (2010), *Physical Chemistry*, 9th edn.

Bacchin, P., Aimar, P. & Field, R. W. (2006), 'Critical and sustainable fluxes: Theory, experiments and applications', *Journal of Membrane Science* **281**(1–2), 42–69.

Bailey, J. E. & Ollis, D. F. (1986), 'Fundamentals of Biochemical engineering'.

Bandyopadhyay, B., Humphrey, A. E. & Taguchi, H. (1967), 'Dynamic measurement of the volumetric oxygen transfer coefficient in fermentation systems', *Biotechnology and Bioengineering* **9**(4), 533–544.

References

Baptista, R. P., Fluri, D. A. & Zandstra, P. W. (2013), 'High density continuous production of murine pluripotent cells in an acoustic perfused bioreactor at different oxygen concentrations.', *Biotechnology and Bioengineering* **110**(2), 648–655.

Barbaroux, M., Gerighausen, S. & Hackel, H. (2014), 'An approach to quality and security of supply for single-use bioreactors', *Advances in Biochemical Engineering/Biotechnology* **138**, 239–272.

Barberis, M., Klipp, E., Vanoni, M. & Alberghina, L. (2007), 'Cell size at S phase initiation: An emergent property of the G1/S network', *PLoS Computational Biology* **3**(4), e64.

Barrett, S., Franklin, J., Stangl, M., Cvetkovic, A. & He, W. (2018), 'Intensification of a multi-product perfusion platform managing growth characteristics at high cell density for maximized volumetric productivity', Cell Culture Engineering XVI.

Barrett, T. A., Wu, A., Zhang, H., Levy, M. S. & Lye, G. J. (2010), 'Microwell engineering characterization for mammalian cell culture process development', *Biotechnology and Bioengineering* **105**(2), 260–275.

Bartholomew, D. J. (2010), 'Principal components analysis', in *International Encyclopedia of Education*, Vol. 2, Elsevier, pp. 374–377.

Bates, R. L., Fondy, P. L. & Fenic, J. G. (1966), 'Impeller characteristics and power', in J. Uh L, VW; Gray, ed., *Mixing: Theory and Practice 1*, Vol. 1, pp. 111–178.

Baur, D., Angarita, M., Müller-Späth, T. & Morbidelli, M. (2015), 'Optimal model-based design of the twin-column CaptureSMB process improves capacity utilization and productivity in protein A affinity capture', *Biotechnology Journal* **11**(1), 135–145.

Baur, D., Angarita, M., Müller-Späth, T., Steinebach, F. & Morbidelli, M. (2016), 'Comparison of batch and continuous multi-column protein A capture processes by optimal design', *Biotechnology Journal* **11**(7), 920–931.

Bausch, M., Schultheiss, C. & Sieck, J. B. (2018), 'Recommendations for comparison of productivity between fed-batch and perfusion processes', *Biotechnology Journal* **14**(2), 1–4.

Beier, S. P. & Jonsson, G. (2009), 'Critical flux determination by flux-stepping', *Wiley InterScience* **56**(7), 1739–1747.

Bendiak, B. & Schachter, H. (1987), 'Control of glycoprotein synthesis: Kinetic mechanism, substrate specificity, and inhibition characteristics of UDP-N-acetylglucosamine:alpha-D-mannoside beta 1-2 N-acetylglucosaminyltransferase II from rat liver', *Journal of Biological Chemistry* **262**(12), 5784–5790.

Benz, G. T. (2011), 'Bioreactor design for chemical engineers', *American Institute of Chemical Engineers* **107**, 21–26.

Beresford, T. P., Fitzsimons, N. A., Brennan, N. L. & Cogan, T. M. (2001), 'Recent advances in cheese microbiology', *International Dairy Journal* **11**, 254–274.

Berg, J. M., Tymoczko, J. L. & Stryer, L. (2007), *Biochemistry*, 6th edn, Sara Tenney.

Berg, P. (1974), 'Potential biohazards of recombinant DNA molecules', *Science* **1114**(1973), 1973–1974.

Berg, P., Baltimore, D., Brenner, S., Roblin, R. & Singer, M. (1975), 'Summary statement of the Asilomar conference on recombinant DNA molecules', *Proceedings of the National Academy of Sciences of the United States of America* **72**, 1981–1984.

Berg, P. & Mertz, J. E. (2010), 'Personal reflections on the origins and emergence of recombinant DNA technology', *Genetics* **184**(1), 9–17.

Berry, B. N., Dobrowsky, T. M., Timson, R. C., Kshirsagar, R., Ryll, T. & Wiltberger, K. (2016), 'Quick generation of Raman spectroscopy based in-process glucose control to influence biopharmaceutical protein product quality during mammalian cell culture', *Biotechnology Progress* **32**(1), 224–234.

References

Bertrand, V., Karst, D. J. & Morbidelli, M. (2019), 'Transcriptome and proteome analysis of steady state in a perfusion CHO cell culture process', *Biotechnology and Bioengineering* pp. 1–14.

Bertrand, V., Vogg, S., Villiger, T. K., Stettler, M., Broly, H., Soos, M. & Morbidelli, M. (2018), 'Proteomic analysis of micro-scale bioreactors as scale-down model for a mAb producing CHO industrial fed-batch platform', *Journal of Biotechnology* **279**, 27–36.

Beyer, B., Schuster, M., Jungbauer, A. & Lingg, N. (2018), 'Microheterogeneity of recombinant antibodies: Analytics and functional impact', *Biotechnology Journal* **13**(1), 1–11.

Bibila, T. A. & Robinson, D. K. (1995), 'In pursuit of the optimal fed-batch process for monoclonal antibody production', *Biotechnology Progress* **11**(1), 1–13.

Bielser, J.-M., Chappuis, L., Xiao, Y., Souquet, J., Broly, H. & Morbidelli, M. (2019a), 'Perfusion cell culture for the production of conjugated recombinant fusion proteins reduces clipping and quality heterogeneity compared to batch-mode processes', *Journal of Biotechnology* **302**, 26–31.

Bielser, J.-M., Domaradzki, J., Souquet, J., Broly, H. & Morbidelli, M. (2019b), 'Semi-continuous scale-down models for clone and operating parameter screening in perfusion bioreactors', *Biotechnology Progress* **35**(3), e2790.

Bielser, J.-M., Wolf, M., Souquet, J., Broly, H. & Morbidelli, M. (2018), 'Perfusion mammalian cell culture for recombinant protein manufacturing: A critical review', *Biotechnology Advances* **36**(4), 1328–1340.

Bödeker, B., Potere, E. & Dove, G. (2013), 'Production of recombinant factor VIII from perfusion cultures: II. Large-scale purification', in R. E. Spier, J. B. Griffiths & W. Berthold, eds, *Animal Cell Technology*, Butterworth-Heinemann, pp. 584–590.

Böhm, E., Voglauer, R., Steinfellner, W., Kunert, R., Borth, N. & Katinger, H. (2004), 'Screening for improved cell performance: Selection of subclones with altered production kinetics or improved stability by cell sorting', *Biotechnology and Bioengineering* **88**(6), 699–706.

Bonham-Carter, J. (2018), 'High productivity harvest – Intensify harvest and displace depth filtration in fed-batch cell culture', in *BioProcess International*.

Bonham-Carter, J. & Shevitz, J. (2011), 'A brief history of perfusion', *BioProcess International* **9**(9), 24–30.

Bosco, B., Paillet, C., Amadeo, I., Mauro, L., Orti, E. & Forno, G. (2017), 'Alternating flow filtration as an alternative to internal spin filter based perfusion process: Impact on productivity and product quality', *Biotechnology Progress* **33**(4), 1–5.

Brányik, T., Vicente, A. A., Dostálek, P. & Teixeira, J. A. (2005), 'Continuous beer fermentation using immobilized yeast cell bioreactor systems', *Biotechnology Progress* **21**(3), 653–663.

Breinlinger, K. J., Hobbs, E. D., Malleo, D., Nevill, J. T. & White, M. P. (2018), 'Movement and selection of micro-objects in a microfluidic apparatus'. U.S. Patent 0099282 A1.

Browne, S. M. & Al-Rubeai, M. (2007), 'Selection methods for high-producing mammalian cell lines', *Trends in Biotechnology* **25**(9), 425–432.

Brühlmann, D., Jordan, M., Hemberger, J., Sauer, M., Stettler, M. & Broly, H. (2015), 'Tailoring recombinant protein quality by rational media design', *Biotechnology Progress* **31**(3), 615–629.

Brühlmann, D., Muhr, A., Parker, R., Vuillemin, T., Bucsella, B., Kalman, F., Torre, S., La Neve, F., Lembo, A., Haas, T., Sauer, M., Souquet, J., Broly, H., Hemberger, J. & Jordan, M. (2017a), 'Cell culture media supplemented with raffinose reproducibly enhances high mannose glycan formation', *Journal of Biotechnology* **252**, 32–42.

Brühlmann, D., Sokolov, M., Butté, A., Sauer, M., Hemberger, J., Souquet, J., Broly, H. & Jordan, M. (2017b), 'Parallel experimental design and multivariate analysis provides efficient screening of cell culture media supplements to improve Biosimilar product quality', *Biotechnology and Bioengineering* **114**(7), 1448–1458.

Brunner, M., Doppler, P., Klein, T., Herwig, C. & Fricke, J. (2018), 'Elevated pCO_2 affects the lactate metabolic shift in CHO cell culture processes', *Engineering in Life Sciences* **18**(3), 204–214.

Buckley, K. & Ryder, A. G. (2017), 'Applications of Raman spectroscopy in biopharmaceutical manufacturing: A short review', *Applied Spectroscopy* **71**(6), 1085–1116.

Bujalski, W., Nienow, A., Chatwin, S. & Cooke, M. (1987), 'The dependency on scale of power numbers of Rushton disc turbines', *Chemical Engineering Science* **42**(2), 317–326.

Bunnak, P., Allmendinger, R., Ramasamy, S. V., Lettieri, P. & Titchener-Hooker, N. J. (2016), 'Life-cycle and cost of goods assessment of fed-batch and perfusion-based manufacturing processes for mAbs', *Biotechnology Progress* **32**(5), 1324–1335.

Cao, D.-S., Xu, Q.-S., Liang, Y.-Z., Chen, X. & Li, H.-D. (2010), 'Prediction of aqueous solubility of druglike organic compounds using partial least squares, back-propagation network and support vector machine', *Journal of Chemometrics* **24**, 584–595.

Caplice, E. & Fitzgerald, G. F. (1999), 'Food fermentations: Role of microorganisms in food production and preservation', *International Journal of Food Microbiology* **50**(1–2), 131–149.

Carrondo, M. J., Alves, P. M., Carinhas, N., Glassey, J., Hesse, F., Merten, O. W., Micheletti, M., Noll, T., Oliveira, R., Reichl, U., Staby, A., Teixeira, A. P., Weichert, H. & Mandenius, C. F. (2012), 'How can measurement, monitoring, modeling and control advance cell culture in industrial biotechnology?', *Biotechnology Journal* **7**(12), 1522–1529.

Carvell, J. P. & Dowd, J. E. (2006), 'On-line measurements and control of viable cell density in cell culture manufacturing processes using radio-frequency impedance', *Cytotechnology* **50**(1–3), 35–48.

Cervera, A. E., Petersen, N., Lantz, A. E., Larsen, A. & Gernaey, K. V. (2009), 'Application of near-infrared spectroscopy for monitoring and control of cell culture and fermentation', *Biotechnology Progress* **25**(6), 1561–1581.

Chakrabarty, A., Buzzard, G. T. & Rundell, A. E. (2013), 'Model-based design of experiments for cellular processes', *Wiley Interdisciplinary Reviews: Systems Biology and Medicine* **5**(2), 181–203.

Chandrasekharan, K. & Calderbank, P. (1981), 'Further observations on the scale-up of aerated mixing vessels', *Chemical Engineering Science* **36**(5), 818–823.

Chen, C., Wong, H. E. & Goudar, C. T. (2018), 'Upstream process intensification and continuous manufacturing', *Current Opinion in Chemical Engineering* **22**, 191–198.

Chen, L., Nguang, S. K., Chen, X. D. & Li, X. M. (2004), 'Modelling and optimization of fed-batch fermentation processes using dynamic neural networks and genetic algorithms', *Biochemical Engineering Journal* **22**(1), 51–61.

Chen, T.-T. (2013), 'Immuno-oncology', *Journal for ImmunoTherapy of Cancer* **1**(18), 1–9.

Chisti, Y. (2000), 'Animal-cell damage in sparged bioreactors', *Trends in Biotechnology* **18**(10), 420–432.

Chisti, Y. (2001), 'Hydrodynamic damage to animal cells', *Critical Reviews in Biotechnology* **21**(2), 67–110.

Chotteau, V. (2017), 'Process development in screening scale bioreactors and perspectives for very high cell density perfusion', *Integrated Continuous Biomanufacturing III*, Suzanne Farid, University College London, United Kingdom Chetan Goudar, Amgen, USA Paula Alves, IBET, Portugal Veena Warikoo, Axcella Health, Inc., USA Eds, ECI Symposium Series. p. 10691.

References

Chu, L. & Robinson, D. K. (2001), 'Industrial choices for protein production by large-scale cell culture', *Current Opinion in Biotechnology* **12**(2), 180–187.

Chugh, P. & Roy, V. (2014), 'Biosimilars: Current scientific and regulatory considerations', *Current Clinical Pharmacology* **9**(1), 53–63.

Chuppa, S., Tsai, Y. S., Yoon, S., Shackleford, S., Rozales, C., Bhat, R., Tsay, G., Matanguihan, C., Konstantinov, K. & Naveh, D. (1997), 'Fermentor temperature as a tool for control of high-density perfusion cultures of mammalian cells', *Biotechnology and Bioengineering* **55**(2), 328–338.

Clincke, M. F., Mölleryd, C., Samani, P. K., Lindskog, E., Fäldt, E., Walsh, K. & Chotteau, V. (2013b), 'Very high density of Chinese hamster ovary cells in perfusion by alternating tangential flow or tangential flow filtration in WAVE bioreactor, Part II: Applications for antibody production and cryopreservation', *Biotechnology Progress* **29**(3), 768–777.

Clincke, M. F., Mölleryd, C., Zhang, Y., Lindskog, E., Walsh, K. & Chotteau, V. (2011), 'Study of a recombinant CHO cell line producing a monoclonal antibody by ATF or TFF external filter perfusion in a WAVE Bioreactor', *BMC Proceedings* **5**(8), 105.

Clincke, M. F., Mölleryd, C., Zhang, Y., Lindskog, E., Walsh, K. & Chotteau, V. (2013a), 'Very high density of CHO cells in perfusion by ATF or TFF in WAVE bioreactor, Part I: Effect of the cell density on the process', *Biotechnology Progress* **29**(3), 754–767.

Coffman, J., Lin, H., Wang, S., Godfrey, S., Orozco, R., Yildirim, S., Salm, J., Hiller, G., Gagnon, M., Farner, R., Kottmeier, B. & Sullivan, D. (2017), 'Balancing continuous, integrated, and batch processing', in *Integrated Continuous Biomanufacturing III*.

Cohen, S. N., Chang, A. C. Y. & Hsu, L. (1972), 'Non-chromosomal antibiotic resistance in bacteria: genetic transformation of *Escherichia coli* by R-factor DNA', *Proceedings of the National Academy of Sciences of the United States of America* **69**, 2110–2114.

Cohen, S. N., Chang, A. C. Y., Boyer, H. W. & Helling, R. B. (1973), 'Construction of biologically functional bacterial plasmids in vitro', *Proceedings of the National Academy of Sciences of the United States of America* **70**(11), 3240–3244.

Colosimo, A., Goncz, K., Holmes, A., Kunzelmann, K., Bennet, M. & Gruenert, D. (2000), 'Transfer and expression of foreign genes in mammalian cells', *BioTechniques* **29**(2), 314–331.

Coronel, J., Klausing, S., Heinrich, C., Noll, T., Figueredo-Cardero, A. & Castilho, L. R. (2016), 'Valeric acid supplementation combined to mild hypothermia increases productivity in CHO cell cultivations', *Biochemical Engineering Journal* **114**, 101–109.

Croughan, M. S., Konstantinov, K. B. & Cooney, C. (2015), 'The future of industrial bioprocessing: Batch or continuous?', *Biotechnology and Bioengineering* **112**(4), 648–651.

Davey, C. L., Davey, H. M., Kell, D. B. & Todd, R. W. (1993), 'Introduction to the dielectric estimation of cellular biomass in real time, with special emphasis on measurements at high volume fractions', *Analytica Chimica Acta* **279**(1), 155–161.

Davis, D., Delia, S., Safc, L., Ross, S., Lyons, D. & Hodzic, I. (2015), 'Modeling perfusion at small scale using ambr© 15', in *ECI Digital Archives*.

De Jesus, M. J., Girard, P., Bourgeois, M., Baumgartner, G., Jacko, B., Amstutz, H. & Wurm, F. M. (2004), 'TubeSpin satellites: A fast track approach for process development with animal cells using shaking technology', *Biochemical Engineering Journal* **17**(3), 217–223.

Demain, A. L. (2007), 'The business of biotechnology', *Industrial Biotechnology* **3**(3), 269–283.

Deschênes, J.-S., Desbiens, A., Perrier, M. & Kamen, A. (2006), 'Use of cell bleed in a high cell density perfusion culture and multivariable control of biomass and metabolite concentrations', *Asia-Pacific Journal of Chemical Engineering* **1**(1–2), 82–91.

References

Deshpande, N. S. & Barigou, M. (1999), 'Performance characteristics of novel mechanical foam breakers in a stirred tank reactor', *Journal of Chemical Technology and Biotechnology* **987**(May), 979–987.

Deshpande, R. R. & Heinzle, E. (2004), 'On-line oxygen uptake rate and culture viability measurement of animal cell culture using microplates with integrated oxygen sensors', *Biotechnology Letters* **26**(9), 763–767.

D'Este, M., Alvarado-Morales, M. & Angelidaki, I. (2017), 'Amino acids production focusing on fermentation technologies: A review', *Biotechnology Advances* **36**(1), 14–25.

Devi, T. T. & Kumar, B. (2017), 'Mass transfer and power characteristics of stirred tank with Rushton and curved blade impeller', *Engineering Science and Technology, an International Journal* **20**(2), 730–737.

Dhir, S., Morrow, K. J., Rhinehart, R. R. & Wiesner, T. (2000), 'Dynamic optimization of hybridoma growth in a fed-batch bioreactor', *Biotechnology and Bioengineering* **67**(2), 197–205.

Dorival-García, N. & Bones, J. (2017), 'Monitoring leachables from single-use bioreactor bags for mammalian cell culture by dispersive liquid-liquid microextraction followed by ultra high performance liquid chromatography quadrupole time of flight mass spectrometry', *Journal of Chromatography A* **1512**, 51–60.

Dowd, J. E., Jubb, A., Kwok, K. E. & Piret, J. M. (2003), 'Optimization and control of perfusion cultures using a viable cell probe and cell specific perfusion rates', *Cytotechnology* **42**(1), 35–45.

Dowd, J. E., Weber, I., Rodriguez, B., Piret, J. M. & Kwok, K. E. (1999), 'Predictive control of hollow-fiber bioreactors for the production of monoclonal antibodies', *Biotechnology and Bioengineering* **63**(4), 484–492.

Du, Z., Treiber, D., Mccarter, J. D., Fomina-Yadlin, D., Saleem, R. A., Mccoy, R. E., Zhang, Y., Tharmalingam, T., Leith, M., Follstad, B. D., Dell, B., Grisim, B., Zupke, C., Heath, C., Morris, A. E. & Reddy, P. (2015), 'Use of a small molecule cell cycle inhibitor to control cell growth and improve specific productivity and product quality of recombinant proteins in CHO cell cultures', *Biotechnology and Bioengineering* **112**(1), 141–155.

Ducommun, P., Bolzonella, I., Marison, I., von Stockar, U. & Rhiel, M. (2002), 'Real-time in situ monitoring of freely suspended and immobilized cell cultures based on mid-infrared spectroscopic measurements', *Biotechnology and Bioengineering* **77**(2), 174–185.

Ducommun, P., Bolzonella, I., Rhiel, M., Pugeaud, P., Von Stockar, U. & Marison, I. W. (2001a), 'On-line determination of animal cell concentration', *Biotechnology and Bioengineering* **72**(5), 515–522.

Duetz, W. A. (2007), 'Microtiter plates as mini-bioreactors: Miniaturization of fermentation methods', *Trends in Microbiology* **15**(10), 469–475.

Ecker, D. M., Jones, S. D. & Levine, H. L. (2015), 'The therapeutic monoclonal antibody market', *mAbs* **7**(1), 9–14.

Eibl, R., Kaiser, S., Lombriser, R. & Eibl, D. (2010), 'Disposable bioreactors: The current state-of-the-art and recommended applications in biotechnology', *Applied Microbiology and Biotechnology* **86**(1), 41–49.

Eleftherios, P. (1991), 'Media additives for protecting freely suspended animal cells against agitation and aeration damage', *Tibtech* **9**, 316–324.

Eon-Duval, A., Gleixner, R., Valax, P., Soos, M., Neunstoecklin, B., Morbidelli, M. & Broly, H. (2013), 'Quality by design applied to a Fc-fusion protein: A case studys', in *Therapeutic Fc-Fusion Proteins*, Wiley-VCH Verlag GmbH and Co. KGaA, Weinheim, pp. 155–189.

Eon-Duval, A., Valax, P., Solacroup, T., Broly, H., Gleixner, R., Strat, C. L. & Sutter, J. (2012), 'Application of the quality by design approach to the drug substance manufacturing process of an Fc fusion protein: Towards a global multiâŁłstep design space', *Journal of Pharmaceutical Sciences* **101**(10), 3604–3618.

Escandar, G. M., Damiani, P. C., Goicoechea, H. C. & Olivieri, A. C. (2006), 'A review of multivariate calibration methods applied to biomedical analysis', *Microchemical Journal* **82**(1), 29–42.

Esmonde-White, K. A., Cuellar, M., Uerpmann, C., Lenain, B. & Lewis, I. R. (2017), 'Raman spectroscopy as a process analytical technology for pharmaceutical manufacturing and bioprocessing', *Analytical and Bioanalytical Chemistry* **409**(3), 637–649.

Feidl, F., Vogg, S., Wolf, M., Podobnik, M., Ruggeri, C., Ulmer, N., ... & Morbidelli, M. (2020). Process-wide control and automation of an integrated continuous manufacturing platform for antibodies. *Biotechnology and Bioengineering*.

Feidl, F. (2019), Digitalization Platform and Supervisory Control of a Continuous Integrated, PhD thesis, ETH Zürich.

Feidl, F., Garbellini, S., Vogg, S., Sokolov, M., Souquet, J., Broly, H., Butté, A. & Morbidelli, M. (2019), 'A new flow cell and chemometric protocol for implementing inâŁłine Raman spectroscopy in chromatography', *Biotechnology Progress* (March), e2847.

Finn, B., Harvey, L. M., McNeil, B. & McNeil, B. (2006), 'Near-infrared spectroscopic monitoring of biomass, glucose, ethanol and protein content in a high cell density baker's yeast fed-batch bioprocess', *Yeast* **23**(7), 507–517.

Fisher, A. C. C., Kamga, M.-H. H., Agarabi, C., Brorson, K., Lee, S. L. L. & Yoon, S. (2018), 'The current scientific and regulatory landscape in advancing integrated continuous biopharmaceutical manufacturing', *Trends in Biotechnology* **37**(3), 253–267.

Fleischaker, R. J. & Sinskey, A. J. (1981), 'Oxygen demand and supply in cell culture', *European Journal of Applied Microbiology and Biotechnology* **12**(4), 193–197.

Fleming, A. (1929), 'On the antibacterial action of cultures of a penicillium, with special reference to their use in the isolation of B. influenzae. 1929', *British Journal of Experimental Pathology* **10**(3), 226–236.

Fogler, H. S. (2008), *Elements of Chemical Reaction Engineering*, 5th edn, Pearson Education.

Frenzel, A., Hust, M. & Schirrmann, T. (2013), 'Expression of recombinant antibodies', *Frontiers in Immunology* **4**(July), 1–20.

Froment, G. F., Bischoff, K. B. & De Wilde, J. (1990), *Chemical Reactor Analysis and Design*, Vol. 2, Wiley New York.

Gagnon, M., Hiller, G., Luan, Y. T., Kittredge, A., Defelice, J. & Drapeau, D. (2011), 'High-end pH-controlled delivery of glucose effectively suppresses lactate accumulation in CHO fed-batch cultures', *Biotechnology and Bioengineering* **108**(6), 1328–1337.

García-Muñoz, S. & Polizzi, M. (2012), 'WSPLS: A new approach towards mixture modeling and accelerated product development', *Chemometrics and Intelligent Laboratory Systems* **114**, 116–121.

Garnier, A., Voyer, R., Tom, R., Perret, S., Jardin, B. & Kamen, A. (1996), 'Dissolved carbon dioxide accumulation in a large scale and high density production of TGFβ receptor with baculovirus infected Sf-9 cells', *Cytotechnology* **22**(1), 53–63.

Gautam, R., Vanga, S., Ariese, F. & Umapathy, S. (2015), 'Review of multidimensional data processing approaches for Raman and infrared spectroscopy', *EPJ Techniques and Instrumentation* **2**(1), 8.

Glassey, J., Gernaey, K. V., Clemens, C., Schulz, T. W., Oliveira, R., Striedner, G. & Mandenius, C.-F. (2011), 'Process analytical technology (PAT) for biopharmaceuticals.', *Biotechnology Journal* **6**(4), 369–377.

Glassey, J. & von Stosch, M., eds (2018), *Hybrid Modeling in Process Industries*, CRC Press.

References

Godawat, R., Konstantinov, K., Rohani, M. & Warikoo, V. (2015), 'End-to-end integrated fully continuous production of recombinant monoclonal antibodies', *Journal of Biotechnology* **213**, 13–19.

Godoy Silva, R., Berdugo, C. & Chalmers, J. J. (2010), 'Aeration, mixing, and hydrodynamics, animal cell bioreactors', in *Encyclopedia of Industrial Biotechnology*, American Cancer Society, pp. 1–27.

Goh, P. S., Ismail, A. F. & Ng, B. C. (2017), 'Raman spectroscopy', *Membrane Characterization* **72**(12), 31–46.

Goletz, S., Stahn, R. & Kreye, S. (2016), Patent WO 2016/193083 A1.

Gomez, N., Ambhaikar, M., Zhang, L., Huang, C.-J. J., Barkhordarian, H., Lull, J. & Gutierrez, C. (2017), 'Analysis of tubespins as a suitable scale-down model of bioreactors for high cell density CHO cell culture', *Biotechnology Progress* **33**(2), 490–499.

Gorenflo, V. M., Angepat, S., Bowen, B. D. & Piret, J. M. (2003), 'Optimization of an acoustic cell filter with a novel air-backflush system', *Biotechnology Progress* **19**(1), 30–36.

Goudar, C., Stevens, J., Le, K., Gupta, S., Tan, C. & Munro, T. (2017), 'Enabling next-generation cell line development using continuous perfusion and nanofluidic technologiese', in *Integrated Continuous Biomanufacturing III*.

Goudar, C. T., Matanguihan, R., Long, E., Cruz, C., Zhang, C., Piret, J. M. & Konstantinov, K. B. (2007), 'Decreased p_{CO2} accumulation by eliminating bicarbonate addition to high cell-density cultures', *Biotechnology and Bioengineering* **96**(6), 1107–1117.

Goudar, C. T., Piret, J. M. & Konstantinov, K. B. (2011), 'Estimating cell specific oxygen uptake and carbon dioxide production rates for mammalian cells in perfusion culture', *Biotechnology Progress* **27**(5), 1347–1357.

Gramer, M. J., Eckblad, J. J., Donahue, R., Brown, J., Shultz, C., Vickerman, K., Priem, P., van den Bremer, E. T., Gerritsen, J. & van Berkel, P. H. (2011), 'Modulation of antibody galactosylation through feeding of uridine, manganese chloride, and galactose', *Biotechnology and Bioengineering* **108**(7), 1591–1602.

Gray, D. R., Chen, S., Howarth, W., Inlow, D. & Maiorella, B. L. (1996), 'CO2 in large-scale and high-density CHO cell perfusion culture', *Cytotechnology* **22**(1–3), 65–78.

Grillberger, L., Kreil, T. R., Nasr, S. & Reiter, M. (2009), 'Emerging trends in plasma-free manufacturing of recombinant protein therapeutics expressed in mammalian cells', *Biotechnology Journal* **4**(2), 186–201.

Gunther, J., Conner, J. & Seborg, D. (2007), 'Fault detection and diagnosis in an industrial fed-batch cell culture process', *Biotechnology Progress* **23**(4), 851–857.

Hammond, M., Marghitoiu, L., Lee, H., Perez, L., Rogers, G., Nashed-Samuel, Y., Nunn, H. & Kline, S. (2014), 'A cytotoxic leachable compound from single-use bioprocess equipment that causes poor cell growth performance', *Biotechnology Progress* **30**(2), 332–337.

Heidemann, R., Lünse, S., Tran, D. & Zhang, C. (2010), 'Characterization of cell-banking parameters for the cryopreservation of mammalian cell lines in 100-mL cryobags', *Biotechnology Progress* **26**(4), 1154–1163.

Heidemann, R., Mered, M., Wang, D. Q., Gardner, B., Zhang, C., Michaels, J., Henzler, H. J., Abbas, N. & Konstantinov, K. (2002), 'A new seed-train expansion method for recombinant mammalian cell lines', *Cytotechnology* **38**(1–3), 99–108.

Helenius, A. & Aebi, M. (2001), 'Intracellular functions of N-linked glycans', *Science* **291**(5512), 2364–2369.

Henry, O., Kwok, E. & Piret, J. M. (2008), 'Simpler non-instrumented batch and semi-continuous cultures provide mammalian cell kinetic data comparable to continuous and perfusion cultures', *Biotechnology Progress* **24**(4), 921–931.

Higel, F., Seidl, A., Sörgel, F. & Friess, W. (2016), 'N-glycosylation heterogeneity and the influence on structure, function and pharmacokinetics of monoclonal antibodies and Fc fusion proteins', *European Journal of Pharmaceutics and Biopharmaceutics* **100**, 94–100.

Hiller, G. W., Ovalle, A. M., Gagnon, M. P., Curran, M. L. & Wang, W. (2017), 'Cell-controlled hybrid perfusion fed-batch CHO cell process provides significant productivity improvement over conventional fed-batch cultures', *Biotechnology and Bioengineering* **114**(7), 1438–1447.

Hoos, A. (2016), 'Development of immuno-oncology drugs: From CTLA4 to PD1 to the next generations', *Nature Reviews Drug Discovery* **15**(4), 235–247.

Hossler, P., Khattak, S. F. & Li, Z. J. (2009), 'Optimal and consistent protein glycosylation in mammalian cell culture'. *Glycobiology* **19**(9), 936–949.

Howard, D. H., Bach, P. B., Berndt, E. R. & Rena, M. C. (2015), 'Pricing in the market for anticancer drugs', *Journal of Economic Perspectives* **29**(1), 1689–1699.

Hsie, A. W., Recio, L., Katz, D. S., Lee, C. Q., Wagner, M. & Schenley, R. L. (1986), 'Evidence for reactive oxygen species inducing mutations in mammalian cells.', *Proceedings of the National Academy of Sciences* **83**(24), 9616–9620.

Hu, W.-S. (2012), *Cell Culture Bioprocess Engineering*. Springer.

Hubert, M. & Engelen, S. (2004), 'Robust PCA and classification in biosciences', *Bioinformatics* **20**(11), 1728–1736.

Hughes, S. S. (2001), 'Making dollars out of DNA: The first major patent in biotechnology and the commercialization of molecular biology, 1974–1980', *Isis* **92**(3), 541–575.

Hughmark, G. A. (1980), 'Power requirements and interfacial area in gas–liquid turbine agitated systems', *Industrial and Engineering Chemistry Process Design and Development* **19**(4), 638–641.

Indahl, U. (2005), 'A twist to partial least squares regression', *Journal of Chemometrics* **19**(1), 32–44.

Ishida, M., Haga, R., Nishimura, N., Matuzaki, H. & Nakano, R. (1990), 'High cell density suspension culture of mammalian anchorage independent cells: Oxygen transfer by gas sparging and defoaming with a hydrophobic net', *Cytotechnology* **4**(3), 215–225.

Ivarsson, M., Noh, H., Morbidelli, M. & Soos, M. (2015), 'Insights into pH-induced metabolic switch by flux balance analysis', *Biotechnology Progress* **31**(2), 347–357.

Jackson, D., Symons, R. H. & Berg, P. (1972), 'Biochemical method for inserting new genetic information into DNA of simian virus 40: Circular DNA molecules containing lambda phage genes and the galactose operon of *Escherichia coli*', *Proceedings of the National Academy of Sciences of the United States of America* **69**, 2904–2909.

Jacquemart, R., Vandersluis, M., Zhao, M., Sukhija, K., Sidhu, N. & Stout, J. (2016), 'A single-use strategy to enable manufacturing of affordable biologics', *Computational and Structural Biotechnology Journal* **14**, 309–318.

Jagschies, G., Lindskog, E., Lacki, K. & Galliher, P. M. (2018), *Biopharmaceutical Processing: Development, Design, and Implementation of Manufacturing Processes*, Elsevier Science.

Jang, J. D. & Barford, J. P. (2000), 'An unstructured kinetic model of macromolecular metabolism in batch and fed-batch cultures of hybridoma cells producing monoclonal antibody', *Biochemical Engineering Journal* **4**(2), 153–168.

Janoschek, S., Schulze, M., Zijlstra, G., Greller, G. & Matuszczyk, J. (2018), 'A protocol to transfer a fed-batch platform process into semi-perfusion mode: The benefit of automated small scale bioreactors compared to shake flasks as scale-down model', *Biotechnology Progress* **35**(2), 8–10.

Jefferis, R. (2005), 'Glycosylation of recombinant antibody therapeutics', *Biotechnology Progress* **21**(1), 11–16.

References

Jefferis, R. (2009), 'Recombinant antibody therapeutics: The impact of glycosylation on mechanisms of action', *Trends in Pharmacological Sciences* **30**(7), 356–362.

Jesus, M. D. & Wurm, F. M. (2011), 'Manufacturing recombinant proteins in kg-ton quantities using animal cells in bioreactors', *European Journal of Pharmaceutics and Biopharmaceutics* **78**(2), 184–188.

Jiang, M., Severson, K. A., Love, J. C., Madden, H., Swann, P., Zang, L. & Braatz, R. D. (2017), 'Opportunities and challenges of real-time release testing in biopharmaceutical manufacturing', *Biotechnology and Bioengineering* **114**(11), 2445–2456.

Jiang, R., Hoesli, N., Mueller, R., Kretz, T., Chen, H., Xu, S. & Bowers, J. (2018), 'Probing lactate metabolism variations in large-scale bioreactors', *Biotechnology Progress* **34**(3), 756–766.

Jimenez del Val, I., Nagy, J. M. & Kontoravdi, C. (2011), 'A dynamic mathematical model for monoclonal antibody N-linked glycosylation and nucleotide sugar donor transport within a maturing Golgi apparatus', *Biotechnology Progress* **27**(6), 1730–1743.

Joao De Jesus, M. & Wurm, F. M. (2013), 'Scale-up and predictability in process development with suspension cultures of mammalian cells for recombinant protein manufacture: comments on a trend reversal', *Pharmaceutical Bioprocessing* **1**(4), 1–3.

Joeris, K., Frerichs, J.-G., Konstantinov, K. & Scheper, T. (2002), 'In-situ microscopy: Online process monitoring of mammalian cell cultures', *Cytotechnology* **38**(1–3), 129–134.

Jolliffe, I. (2005), 'Principal component analysis', in *Encyclopedia of Statistics in Behavioral Science*, John Wiley and Sons.

Jordan, M. & Jenkins, N. (2007), 'Tools for high-throughput medium and process optimization', *Methods in Biotechnology* **24**, 193–202.

Jordan, M., Kinnon, N. M., Monchois, V., Stettler, M. & Broly, H. (2018), 'Intensification of large-scale cell culture processes', *Current Opinion in Chemical Engineering* **22**, 253–257.

Jordan, M., Voisard, D., Berthoud, A. & Tercier, L. (2012), 'Cell culture medium improvement by rigorous shuffling of components using media blending', *Cytotechnology* **65**(1), 31–40.

Karst, D. J., Scibona, E., Serra, E., Bielser, J.-M. M., Souquet, J., Stettler, M., Broly, H., Soos, M., Morbidelli, M. & Villiger, T. K. (2017b), 'Modulation and modeling of monoclonal antibody N-linked glycosylation in mammalian cell perfusion reactors', *Biotechnology and Bioengineering* **114**(9), 1–37.

Karst, D. J., Serra, E., Villiger, T. K., Soos, M. & Morbidelli, M. (2016), 'Characterization and comparison of ATF and TFF in stirred bioreactors for continuous mammalian cell culture processes', *Biochemical Engineering Journal* **110**, 17–26.

Karst, D. J., Steinebach, F. & Morbidelli, M. (2018), 'Continuous integrated manufacturing of therapeutic proteins', *Current Opinion in Biotechnology* **53**, 76–84.

Karst, D. J., Steinebach, F., Soos, M. & Morbidelli, M. (2017a), 'Process performance and product quality in an integrated continuous antibody production process', *Biotechnology and Bioengineering* **114**(2), 298–307.

Karst, D. J., Steinhoff, R. F., Kopp, M. R. G., Serra, E., Soos, M., Zenobi, R. & Morbidelli, M. (2017c), 'Intracellular CHO cell metabolite profiling reveals steady-state dependent metabolic fingerprints in perfusion culture', *Biotechnology Progress* **33**(4), 879–890.

Karst, D. J., Steinhoff, R. F., Kopp, M. R., Soos, M., Zenobi, R. & Morbidelli, M. (2017d), 'Isotope labeling to determine the dynamics of metabolic response in CHO cell perfusion bioreactors using MALDI-TOF-MS', *Biotechnology Progress* **33**(6), 1630–1639.

Kaufmann, H., Mazur, X., Fussenegger, M. & Bailey, J. E. (1999), 'Influence of low temperature on productivity, proteome and protein phosphorylation of CHO cells', *Biotechnology and Bioengineering* **63**(5), 573–582.

References

Kawase, Y., Halard, B. & Moo-Young, M. (1992), 'Liquid-phase mass transfer coefficients in bioreactors', *Biotechnology and Bioengineering* **39**(11), 1133–1140.

Kelly, P. S., McSweeney, S., Coleman, O., Carillo, S., Henry, M., Chandran, D., Kellett, A., Bones, J., Clynes, M., Meleady, P. & Barron, N. (2016), 'Process-relevant concentrations of the leachable bDtBPP impact negatively on CHO cell production characteristics', *Biotechnology Progress* **32**(6), 1547–1558.

Kelly, W. J. (2008), 'Using computational fluid dynamics to characterize and improve bioreactor performance', *Biotechnology and Applied Biochemistry* **49**(4), 225.

Kelly, W., Scully, J., Zhang, D., Feng, G., Lavengood, M., Condon, J., Knighton, J. & Bhatia, R. (2014), 'Understanding and modeling alternating tangential flow filtration for perfusion cell culture', *Biotechnology Progress* **30**(6), 1291–1300.

Kettaneh, N., Berglund, A. & Wold, S. (2005), 'PCA and PLS with very large data sets', *Computational Statistics and Data Analysis* **48**(1), 69–85.

Khawli, L. A., Goswami, S., Hutchinson, R., Kwong, Z. W., Yang, J., Wang, X., Yao, Z., Sreedhara, A., Cano, T., Tesar, D., Nijem, I., Allison, D. E., Wong, P. Y., Kao, Y. H., Quan, C., Joshi, A., Harris, R. J. & Motchnik, P. (2010), 'Charge variants in IgG1: Isolation, characterization, in vitro binding properties and pharmacokinetics in rats', *mAbs* **2**(6), 613–624.

Kimura, R. & Miller, W. M. (1996), 'Effects of elevated pCO(2) and/or osmolality on the growth and recombinant tPA production of CHO cells.', *Biotechnology and Bioengineering* **52**(1), 152–160.

Kiparissides, A., Koutinas, M., Kontoravdi, C., Mantalaris, A. & Pistikopoulos, E. N. (2011), 'Closing the loop' in biological systems modeling: From the in silico to the in vitro', *Automatica* **47**(6), 1147–1155.

Kirdar, A. O., Chen, G., Weidner, J. & Rathore, A. S. (2010), 'Application of near-infrared (NIR) spectroscopy for screening of raw materials used in the cell culture medium for the production of a recombinant therapeutic protein', *Biotechnology Progress* **26**(2), 527–531.

Kiviharju, K., Salonen, K., Moilanen, U. & Eerikäinen, T. (2008), 'Biomass measurement online: The performance of in situ measurements and software sensors', *Journal of Industrial Microbiology and Biotechnology* **35**(7), 657–665.

Klutz, S., Holtmann, L., Lobedann, M. & Schembecker, G. (2016), 'Cost evaluation of antibody production processes in different operation modes', *Chemical Engineering Science* **141**, 63–74.

Klutz, S., Magnus, J., Lobedann, M., Schwan, P., Maiser, B., Niklas, J., Temming, M. & Schembecker, G. (2015), 'Developing the biofacility of the future based on continuous processing and single-use technology', *Journal of Biotechnology* **213**, 120–130.

Kochanowski, N., Blanchard, F., Cacan, R., Chirat, F., Guedon, E., Marc, A. & Goergen, J. L. (2008), 'Influence of intracellular nucleotide and nucleotide sugar contents on recombinant interferon-γ glycosylation during batch and fed-batch cultures of CHO cells', *Biotechnology and Bioengineering* **100**(4), 721–733.

Kohrt, H. E., Tumeh, P. C., Benson, D., Bhardwaj, N., Brody, J., Formenti, S., Fox, B. A., Galon, J., June, C. H., Kalos, M., Kirsch, I., Kleen, T., Kroemer, G., Lanier, L., Levy, R., Lyerly, H. K., Maecker, H., Marabelle, A., Melenhorst, J., Miller, J., Melero, I., Odunsi, K., Palucka, K., Peoples, G., Ribas, A., Robins, H., Robinson, W., Serafini, T., Sondel, P., Vivier, E., Weber, J., Wolchok, J., Zitvogel, L., Disis, M. L. & Cheever, M. A. (2016), 'Immunodynamics: A cancer immunotherapy trials network review of immune monitoring in immuno-oncology clinical trials', *Journal for ImmunoTherapy of Cancer* **4**(1), 1–16.

References

Kolwyck, D., Mcsweeney, M. & Johns, J. (2017), Biomanufacturing Technology Roadmap: Supply Partnership Management, technical report, BioPhorum Operations Group Ltd.

Konstantinov, K. B. & Cooney, C. L. (2015), 'White paper on continuous bioprocessing May 20–21, 2014 continuous manufacturing symposium', *Journal of Pharmaceutical Sciences* **104**(3), 813–820.

Konstantinov, K. B., Tsai Yeong-shou, Moles, D. & Matanguihan, R. (1996), 'Control of long-term perfusion Chinese hamster ovary cell culture by glucose auxostat', *Biotechnology Progress* **12**(1), 100–109.

Konstantinov, K., Chuppa, S., Sajan, E., Tsai, Y., Yoon, S. & Golini, F. (1994), 'Real-time biomass-concentration monitoring in animal-cell cultures', *Trends in Biotechnology* **12**(8), 324–333.

Konstantinov, K., Goudar, C., Ng, M., Meneses, R., Thrift, J., Chuppa, S., Matanguihan, C., Michaels, J. & Naveh, D. (2006), 'The "push-to-low" approach for optimization of high-density perfusion cultures of animal cells', *Advances in Biochemical Engineering/Biotechnology* **101**(July), 75–98.

Kontoravdi, C., Pistikopoulos, E. N. & Mantalaris, A. (2010), 'Systematic development of predictive mathematical models for animal cell cultures', *Computers and Chemical Engineering* **34**(8), 1192–1198.

Krambeck, F. J. & Betenbaugh, M. J. (2005), 'A mathematical model of N-linked glycosylation', *Biotechnology and Bioengineering* **92**(6), 711–728.

Krättli, M., Müller-Späth, T. & Morbidelli, M. (2013), 'Multifraction separation in countercurrent chromatography (MCSGP).', *Biotechnology and Bioengineering* **110**(9), 2436–2444.

Kratzer, R., Dorn, I., Mcnaull, S., Rode, C., Lilly, E., Shea, L. O., Campbell, C. & Diluzio, W. (2017), Biomanufacturing Technology Roadmap: Process Technologies, technical report, BioPhorum Operations Group Ltd.

Kreye, S., Stahn, R., Nawrath, K., Danielczyk, A., Goletz, S. & Gmbh, G. (2015), 'GlycoExpress: A toolbox for the high yield production of glycooptimized fully human biopharmaceuticals in perfusion bioreactors at different scales', in *ECI Digital Archives*.

Kumar, N., Bansal, A., Sarma, G. & Rawal, R. K. (2014), 'Chemometrics tools used in analytical chemistry: An overview', *Talanta* **123**, 186–199.

Kunkel, J. P., Jan, D. C., Butler, M. & Jamieson, J. C. (2000), 'Comparisons of the glycosylation of a monoclonal antibody produced under nominally identical cell culture conditions in two different bioreactors', *Biotechnology Progress* **16**(3), 462–470.

Kunkel, J. P., Jan, D. C., Jamieson, J. C. & Butler, M. (1998), 'Dissolved oxygen concentration in serum-free continuous culture affects N-linked glycosylation of a monoclonal antibody', *Journal of Biotechnology* **62**, 55–71.

Lai, T., Yang, Y. & Ng, S. K. (2013), 'Advances in mammalian cell line development technologies', *Pharmaceuticals* **6**, 579–603.

Lairson, L., Henrissat, B., Davies, G. & Withers, S. (2008), 'Glycosyltransferases: Structures, functions, and mechanisms', *Annual Review of Biochemistry* **77**(1), 521–555.

Lameris, R., de Bruin, R. C., Schneiders, F. L., van Bergen en Henegouwen, P. M., Verheul, H. M., de Gruijl, T. D. & van der Vliet, H. J. (2014), 'Bispecific antibody platforms for cancer immunotherapy', *Critical Reviews in Oncology/Hematology* **92**(3), 153–165.

Langer, E. S. (2011), 'Trends in perfusion bioreactors: The next revolution in bioprocessing?', *BioProcess International* **9**(10), 18–22.

Langer, E. S. (2014), 'Continuous bioprocessing and perfusion: Wider adoption coming as bioprocessing matures', *BioProcessing Journal* **13**(1), 43–49.

References

Lavery, M. & Nienow, A. W. (1987), 'Oxygen transfer in animal cell culture medium', *Biotechnology and Bioengineering* **30**(3), 368–373.

Le, K., Tan, C., Gupta, S., Guhan, T., Barkhordarian, H., Lull, J., Stevens, J. & Munro, T. (2018), 'A novel mammalian cell line development platform utilizing nanofluidics and optoelectro positioning technology', *Biotechnology Progress* **34**(6), 1438–1446.

Li, J., Wong, C. L., Vijayasankaran, N., Hudson, T. & Amanullah, A. (2012), 'Feeding lactate for CHO cell culture processes: Impact on culture metabolism and performance', *Biotechnology and Bioengineering* **109**(5), 1173–1186.

Lim, Y., Wong, N. S. C., Lee, Y. Y., Ku, S. C. Y., Wong, D. C. F. & Yap, M. G. S. (2010), 'Engineering mammalian cells in bioprocessing: Current achievements and future perspectives', *Biotechnology and Applied Biochemistry* **55**(4),175–189.

Lin, H., Leighty, R. W., Godfrey, S. & Wang, S. B. (2017), 'Principles and approach to developing mammalian cell culture media for high cell density perfusion process leveraging established fed-batch media', *Biotechnology Progress* **33**(4), 891–901.

Ling, W. L. (2015), 'Development of protein-free medium for therapeutic protein production in mammalian cells: Recent advances and perspectives', *Pharmaceutical Bioprocessing* **3**, 215–226.

Lobban, P. (1972), An Enzymatic Method for End-to-End Joining of DNA Molecules, PhD thesis, Stanford University.

Lobban, P. & Kaiser, A. (1973), 'Enzymatic end-to-end joining of DNA molecules', *Journal of Molecular Biology* **78**, 483–471.

Löffelholz, C., Kaiser, S. C., Kraume, M., Eibl, R. & Eibl, D. (2014), 'Dynamic single-use bioreactors used in modern liter- and m3- scale biotechnological processes: Engineering characteristics and scaling up', *Advances in Biochemical Engineering/Biotechnology* **138**, 1–44.

Long, Q., Liu, X., Yang, Y., Li, L., Harvey, L., McNeil, B. & Bai, Z. (2014), 'The development and application of high throughput cultivation technology in bioprocess development', *Journal of Biotechnology* **192**, 323–338.

Losfeld, M.-E., Scibona, E., Lin, C.-W., Villiger, T. K., Gauss, R., Morbidelli, M. & Aebi, M. (2017), 'Influence of protein/glycan interaction on site-specific glycan heterogeneity', *The FASEB Journal* **31**(10), 4623–4635.

Luo, Y. & Chen, G. (2007), 'Combined approach of NMR and chemometrics for screening peptones used in the cell culture medium for the production of a recombinant therapeutic protein', *Biotechnology and Bioengineering* **97**(6), 1654–1659.

Luttmann, R., Borchert, S. O., Mueller, C., Loegering, K., Aupert, F., Weyand, S., Kober, C., Faber, B. & Cornelissen, G. (2015), 'Sequential/parallel production of potential Malaria vaccines: A direct way from single batch to quasi-continuous integrated production', *Journal of Biotechnology* **213**, 83–96.

Luttmann, R., Bracewell, D. G., Cornelissen, G., Gernaey, K. V., Glassey, J., Hass, V. C., Kaiser, C., Preusse, C., Striedner, G. & Mandenius, C. (2012), 'Soft sensors in bioprocessing: A status report and recommendations', *Biotechnology Journal* **7**(8), 1040–1048.

Ma, N., Mollet, M. & Chalmers, J. J. (2003), 'Aeration, mixing and hydrodynamics in bioreactors', in *Encyclopedia of Cell Technology*, John Wiley and Sons, Inc., Hoboken, pp. 225–248.

Manchester, K. L. (2007), 'Louis Pasteur, fermentation, and a rival', *South African Journal of Science* **103**(9–10), 377–380.

Marks, D. M. (2003), 'Equipment design considerations for large scale cell culture', *Cytotechnology* **42**(1), 21–33.

References

Matthews, T. E., Berry, B. N., Smelko, J., Moretto, J., Moore, B. & Wiltberger, K. (2016), 'Closed loop control of lactate concentration in mammalian cell culture by Raman spectroscopy leads to improved cell density, viability, and biopharmaceutical protein production', *Biotechnology and Bioengineering* **113**(11), 2416–2424.

McCoy, R. E., Costa, N. a. & Morris, A. E. (2015), 'Factors that determine stability of highly concentrated chemically defined production media', *Biotechnology Progress* **31**(2), 493–502.

McCracken, N. A., Kowle, R. & Ouyang, A. (2014), 'Control of galactosylated glycoforms distribution in cell culture system', *Biotechnology Progress* **30**(3), 547–553.

McGovern, P. E., Glusker, D. L., Exner, L. J. & Voigt, M. M. (1996), 'Neolithic resinated wine', *Nature* **381**, 480–481.

McGovern, P. E., Zhang, J., Tang, J., Zhang, Z., Hall, G. R., Moreau, R. A., Nunez, A., Butrym, E. D., Richards, M. P., Wang, C.-S., Cheng, G., Zhao, Z. & Wang, C. (2004), 'Fermented beverages of pre- and proto-historic China', *Proceedings of the National Academy of Sciences* **101**(51), 17593–17598.

Meglen, R. R. (1992), 'Examining large databases: A chemometric approach using principal component analysis', *Marine Chemistry* **39**(1–3), 217–237.

Mehdizadeh, H., Lauri, D., Karry, K. M., Moshgbar, M., Procopio-Melino, R. & Drapeau, D. (2015), 'Generic Raman-based calibration models enabling real-time monitoring of cell culture bioreactors', *Biotechnology Progress* **31**(4), 1004–1013.

Meier, S. J., Hatton, T. A. & Wang, D. I. (1999), 'Cell death from bursting bubbles: Role of cell attachment to rising bubbles in sparged reactors', *Biotechnology and Bioengineering* **62**(4), 468–478.

Mercier, S. M., Diepenbroek, B., Wijffels, R. H. & Streefland, M. (2014), 'Multivariate PAT solutions for biopharmaceutical cultivation: current progress and limitations', *Trends in Biotechnology* **32**(6), 329–336.

Mercier, S. M., Rouel, P. M. & Lebrun, P. (2016), 'Process analytical technology tools for perfusion cell culture', *Engineering in Life Sciences* **16**(1), 25–35.

Mercille, S., Johnson, M., Lanthier, S., Kamen, A. A. & Messie, B. (2000), 'Understanding factors that limit the productivity of suspension-based perfusion cultures operated at high medium renewal rates', *Biotechnology and Bioengineering* **67**(4), 435–450.

Mertz, J. E. & Davis, R. (1972), 'Cleavage of DNA by RI restriction endonuclease generates cohesive ends', *Proceedings of the National Academy of Sciences of the United States of America* **69**(11), 3370–3374.

Meuwly, F., Weber, U., Ziegler, T., Gervais, A., Mastrangeli, R., Crisci, C., Rossi, M., Bernard, A., von Stockar, U. & Kadouri, A. (2006), 'Conversion of a CHO cell culture process from perfusion to fed-batch technology without altering product quality', *Journal of Biotechnology* **123**(1), 106–116.

Miller, W. M., Blanch, H. W. & Wilke, C. R. (1988), 'A kinetic analysis of hybridoma growth and metabolism in batch and continuous suspension culture: effect of nutrient concentration, dilution rate, and pH', *Biotechnology and Bioengineering* **32**(8), 947–965.

Miller, W. M., Wilke, C. R. & Blanch, H. W. (1987), 'Effects of dissolved oxygen concentration on hybridoma growth and metabolism in continuous culture', *Journal of Cellular Physiology* **132**(3), 524–530.

Mokuolu, S. (2018), 'New standards define single-use materials qualification', *Pharmaceutical Technology* **42**(2), 52–53.

Monteil, D. T., Juvet, V., Paz, J., Moniatte, M., Baldi, L., Hacker, D. L. & Wurm, F. M. (2016), 'A comparison of orbitally-shaken and stirred-tank bioreactors: pH modulation and bioreactor type affect CHO cell growth and protein glycosylation', *Biotechnology Progress* **32**(5), 1174–1180.

References

Morari, M. & Zafiriou, E. (1989), *Robust Process Control.* Prentice Hall.

Morrow, J., Cohen, S. N., Chang, A. C., Boyer, H. W. & Goodman, H. M. (1974), 'Replication and transcription of eukaryotic DNA in *Escherichia coli*', *Proceedings of the National Academy of Sciences of the United States of America* **71**(5), 1743–1747.

Moussa, A. S., Soos, M., Sefcik, J. & Morbidelli, M. (2007), 'Effect of solid volume fraction on aggregation and breakage in colloidal suspensions in batch and continuous stirred tanks', *Langmuir* **23**(4), 1664–1673.

Moyle, D. (2017), Biomanufacturing Technology Roadmap: Modular and Mobile, Technical report, BioPhorum Operations Group Ltd.

Mulukutla, B. C., Gramer, M. & Hu, W. S. (2012), 'On metabolic shift to lactate consumption in fed-batch culture of mammalian cells', *Metabolic Engineering* **14**(2), 138–149.

Narayanan, H., Luna, M., von Stoch, M., Cruz Bournazou, M., Polotti, G., Morbidelli, M., Butté, A. & Sokolov, M. (2019a), 'Bioprocess in the digital age: The role of process models', *Biotechnology Journal* **15**(1), https://doi.org/10.1002/biot.201900172.

Narayanan, H., Sokolov, M., Butté, A. & Morbidelli, M. (2019b), 'Decision Tree-PLS (DT-PLS) algorithm for the development of process: Specific local prediction models', *Biotechnology Progress* **35**(4), e2818.

Narayanan, H., Sokolov, M., Morbidelli, M. & Butté, A. (2019c), 'A new generation of predictive models: The added value of hybrid models for manufacturing processes of therapeutic proteins', *Biotechnology and Bioengineering* **116**(10), 2540–2549.

Nasr, M. M., Krumme, M., Matsuda, Y., Trout, B. L., Badman, C., Mascia, S., Cooney, C. L., Jensen, K. D., Florence, A., Johnston, C., Konstantinov, K. & Lee, S. L. (2017), 'Regulatory perspectives on continuous pharmaceutical manufacturing: Moving from theory to practice, September 26–27, 2016, International Symposium on the Continuous Manufacturing of Pharmaceuticals', *Journal of Pharmaceutical Sciences* **106**(11), 3199–3206.

Neunstoecklin, B., Stettler, M., Solacroup, T., Broly, H., Morbidelli, M. & Soos, M. (2015), 'Determination of the maximum operating range of hydrodynamic stress in mammalian cell culture', *Journal of Biotechnology* **194**, 100–109.

Neunstoecklin, B., Villiger, T. K., Lucas, E., Stettler, M., Broly, H., Morbidelli, M. & Soos, M. (2016), 'Pilot-scale verification of maximum tolerable hydrodynamic stress for mammalian cell culture', *Applied Microbiology and Biotechnology* **100**(8), 3489–3498.

Nienow, A. W. (1997), 'On impeller circulation and mixing effectiveness in the turbulent flow regime', *Chemical Engineering Science* **52**(15), 2557–2565.

Nienow, A. W. (1998), 'Hydrodynamics of stirred bioreactors', *Applied Mechanics Reviews* **51**(1), 3–32.

Nienow, A. W. (2006), 'Reactor engineering in large scale animal cell culture', *Cytotechnology* **50**(1–3), 9–33.

Nienow, A. W. (2010), 'Impeller selection for animal cell culture', in *Encyclopedia of Industrial Biotechnology*, American Cancer Society, pp. 1–25.

Nienow, A. W., Rielly, C. D., Brosnan, K., Bargh, N., Lee, K., Coopman, K. & Hewitt, C. J. (2013), 'The physical characterisation of a microscale parallel bioreactor platform with an industrial CHO cell line expressing an IgG4', *Biochemical Engineering Journal* **76**, 25–36.

Opel, C. F., Li, J. & Amanullah, A. (2010), 'Quantitative modeling of viable cell density, cell size, intracellular conductivity, and membrane capacitance in batch and fed-batch CHO processes using dielectric spectroscopy', *Biotechnology Progress* **26**(4), 1187–1199.

Ozturk, S. S. (1996), 'Engineering challenges in high density cell culture systems', *Cytotechnology* **22**, 3–16.

References

Ozturk, S. S. (2014), 'Opportunities and challenges for the implementation of continuous processing in biomanufacturing', in G. Subramanian, ed., *Continuous Processing in Pharmaceutical Manufacturing*, Wiley-Blackwell, chapter 18, pp. 457–478.

Ozturk, S. S. & Hu, W.-S. (2006), *Cell Culture Technology for Pharmaceutical and Cell-Based Therapies*, Taylor and Francis.

Ozturk, S. S. & Kompala, D. S. (2006), 'Optimization of high cell density perfusion bioreactors', in *Cell Culture Technology for Pharmaceutical and Cell-Based Therapies*, Taylor and Francis Group, Boca Raton, pp. 387–416.

Ozturk, S. S. & Palsson, B. O. (1990), 'Effects of dissolved oxygen on hybridoma cell growth, metabolism, and antibody production kinetics in continuous culture', *Biotechnology Progress* **6**(6), 437–446.

Ozturk, S. S. & Palsson, B. O. (1991), 'Growth, metabolic, and antibody production kinetics of hybridoma cell culture: 2. Effects of serum concentration, dissolved oxygen concentration, and medium pH in a batch reactor', *Biotechnology Progress* **7**(6), 481–494.

Pardee, A. B. (1989), 'G1 events and regulation of cell proliferation', *Science* **246**(4930), 603–608.

Park, J. H., Noh, S. M., Woo, J. R., Kim, J. W. & Lee, G. M. (2015), 'Valeric acid induces cell cycle arrest at G1 phase in CHO cell cultures and improves recombinant antibody productivity', *Biotechnology Journal* **11**(4), 487–496.

Pasteur, L. (1885), Mémoire sur la fermentation alcoolique, PhD thesis, Académie des sciences.

Pattison, R., Swamy, J., Mendenhall, B., Hwang, C. & Frohlich, B. (2000), 'Measurement and control of dissolved carbon dioxide in mammalian cell culture processes using an in situ fiber optic chemical sensor', *Biotechnology Progress* **16**(5), 769–774.

Pfister, D., Nicoud, L. & Morbidelli, M. (2018), *Continuous Biopharmaceutical Processes*, Cambridge University Press.

Pham, P. L., Kamen, A. & Durocher, Y. (2006), 'Large-scale transfection of mammalian cells for the fast production of recombinant protein', *Molecular Biotechnology* **34**(2), 225–237.

Pilkington, P. H., Margaritis, A., Mensour, N. A. & Russell, I. (1998), 'Fundamentals of immobilised yeast cells for continuous beer fermentation: A review', *Journal of the Institute of Brewing* **104**(1), 19–31.

Pohlscheidt, M., Jacobs, M., Wolf, S., Thiele, J., Jockwer, A., Gabelsberger, J., Jenzsch, M., Tebbe, H. & Burg, J. (2013), 'Optimizing capacity utilization by large scale 3000 L perfusion in seed train bioreactors', *Biotechnology Progress* **29**(1), 222–229.

Politis, S., Colombo, P., Colombo, G. & Rekkas, D. (2017), 'Design of experiments (DoE) in pharmaceutical development', *Drug Development and Industrial Pharmacy* **43**(6), 889–901.

Pollock, J., Coffman, J., Ho, S. V. & Farid, S. S. (2017), 'Integrated continuous bioprocessing: Economic, operational, and environmental feasibility for clinical and commercial antibody manufacture', *Biotechnology Progress* **33**(4), 854–866.

Pollock, J., Ho, S. V. & Farid, S. S. (2013), 'Fed-batch and perfusion culture processes: Economic, environmental, and operational feasibility under uncertainty', *Biotechnology and Bioengineering* **110**(1), 206–219.

Pörtner, R. (2015), 'Bioreactors for mammalian cells', in M. Al-Rubeai, ed., *Animal Cell Culture*, Springer International Publishing, pp. 89–135.

Pörtner, R. & Schäfer, T. (1996), 'Modelling hybridoma cell growth and metabolism a comparison of selected models and data', *Journal of Biotechnology* **49**(1–3), 119–135.

Radoniqi, F., Zhang, H., Bardliving, C. L., Shamlou, P. & Coffman, J. (2018), 'Computational fluid dynamic modeling of alternating tangential flow filtration for perfusion cell culture', *Biotechnology and Bioengineering* **115**(11), 2751–2759.

References

Raghunath, B., Bin, W., Pattnaik, P. & Janssens, J. (2013), 'Best practices for optimization and scale-up of microfiltration TFF processes', *BioProcessing Journal* **11**(1), 30–40.

Ramakrishnan, B., Boeggeman, E., Ramasamy, V. & Qasba, P. K. (2004), 'Structure and catalytic cycle of β-1,4-galactosyltransferase', *Current Opinion in Structural Biology* **14**(5), 593–600.

Ranganathan, P. & Sivaraman, S. (2011), 'Investigations on hydrodynamics and mass transfer in gas–liquid stirred reactor using computational fluid dynamics', *Chemical Engineering Science* **66**(14), 3108–3124.

Rathore, A. S. (2009), 'Roadmap for implementation of quality by design (QbD) for biotechnology products'. *Trend in Biotechnology*, **27**(9), 546–553, https://doi.org/10.1016/j.tibtech.2009.06.006.

Rathore, A. S. (2014), 'QbD/PAT for bioprocessing: Moving from theory to implementation'. *Current Opinion in Chemical Engineering*, **6**, 1–8, https://doi.org/10.1016/j.coche.2014.05.006.

Rathore, A. S., Kateja, N. & Kumar, D. (2018), 'Process integration and control in continuous bioprocessing', *Current Opinion in Chemical Engineering* **22**, 18–25.

Rathore, A. S., Pathak, M. & Godara, A. (2016), 'Process development in the QbD paradigm: Role of process integration in process optimization for production of biotherapeutics', *Biotechnology Progress* **32**(2), 355–362.

Rathore, A. S. & Winkle, H. (2009), 'Quality by design for biopharmaceuticals', *Nature* **27**(1), 26–34.

Read, E., Park, J., Shah, R., Riley, B., Brorson, K. & Rathore, A. (2010a), 'Process analytical technology (PAT) for biopharmaceutical products: Part I. Concepts and applications', *Biotechnology and Bioengineering* **105**(2), 276–284.

Read, E., Shah, R., Riley, B., Park, J., Brorson, K. & Rathore, A. (2010b), 'Process analytical technology (PAT) for biopharmaceutical products: Part II. Concepts and applications', *Biotechnology and Bioengineering* **105**(2), 285–295.

Reinhart, D., Damjanovic, L., Kaisermayer, C. & Kunert, R. (2015), 'Benchmarking of commercially available CHO cell culture media for antibody production', *Applied Microbiology and Biotechnology* **99**(11), 4645–4657.

Rivinoja, A., Hassinen, A., Kokkonen, N., Kauppila, A. & Kellokumpu, S. (2009), 'Elevated Golgi pH impairs terminal N-glycosylation by inducing mislocalization of Golgi glycosyltransferases', *Journal of Cellular Physiology* **220**(1), 144–154.

Rodrigues, M. E., Costa, A. R., Henriques, M., Azeredo, J. & Oliveira, R. (2010), 'Technological progresses in monoclonal antibody production systems', *Biotechnology Progress* **26**(2), 332–351.

Rouiller, Y., Bielser, J.-M., Brühlmann, D., Jordan, M., Broly, H. & Stettler, M. (2016), 'Screening and assessment of performance and molecule quality attributes of industrial cell lines across different fed-batch systems', *Biotechnology Progress* **32**(1), 160–170.

Rouiller, Y., Périlleux, A., Collet, N., Jordan, M., Stettler, M. & Broly, H. (2013), 'A high-throughput media design approach for high performance mammalian fed-batch cultures', *mAbs* **5**(3), 501–511.

Rouiller, Y., Solacroup, T., Deparis, V., Barbafieri, M., Gleixner, R., Broly, H. & Eon-Duval, A. (2012), 'Application of quality by design to the characterization of the cell culture process of an Fc-fusion protein', *European Journal of Pharmaceutics and Biopharmaceutics* **81**(2), 426–437.

Routledge, S. J. (2012), 'Beyond de-foaming: The effects of antifoams on bioprocess productivity', *Computational and Structural Biotechnology Journal* **3**(4), e201210001.

References

Roy, J. (2009), 'Glycosylation of antibody therapeutics: Optimisation for purpose', in *Methods in Molecular Biology*, Vol. 483, pp. 223–238.

Roychoudhury, P., O'Kennedy, R., McNeil, B. & Harvey, L. M. (2007), *Analytica Chimica Acta* **590**(1), 110–117.

Running, J. A. & Bansal, K. (2016), 'Oxygen transfer rates in shaken culture vessels from Fernbach flasks to microtiter plates', *Biotechnology and Bioengineering* **113**(8), 1729–1735.

Rutherford, K., Mahmoudi, S. M. S., Lee, K. C. & Yianneskis, M. (1996), 'The influence of Rushton impeller blade and disk thickness on the mixing characteristics of stirred vessels', *Chemical Engineering Research and Design* **74**(3), 369–378.

Ryan, P. W., Li, B., Shanahan, M., Leister, K. J. & Ryder, A. G. (2010), 'Prediction of cell culture media performance using fluorescence spectroscopy', *Analytical Chemistry* **82**(4), 1311–1317.

Saha, D., Soos, M., Lüthi, B., Holzner, M., Liberzon, A., Babler, M. U. & Kinzelbach, W. (2014), 'Experimental characterization of breakage rate of colloidal aggregates in axisymmetric extensional flow', *Langmuir* **30**(48), 14385–14395.

Sajjadi, S. & Yianneskis, M. (2003), 'Semibatch emulsion polymerization of methyl methacrylate with a neat monomer feed', *Polymer Reaction Engineering* **11**(4), 715–736.

Sano, C. (2009), 'History of glutamate production', *American Journal of Clinical Nutrition* **90**(3), 728–732.

Sawyer, D., Sanderson, K., Lu, R., Daszkowski, T., Clark, E., Mcduff, P., Astrom, J., Heffernan, C., Duffy, L., Poole, S., Ryll, T., Sheehy, P., Strachan, D., Souquet, J., Beattie, D., Pollard, D., Stauch, O., Bezy, P., Sauer, T., Boettcher, L., Simpson, C., Dakin, J., Pitt, S. & Boyle, A. (2017a), Biomanufacturing Technology Roadmap: Overview, Technical report, BioPhorum Operations Group Ltd.

Sawyer, D., Sanderson, K., Lu, R., Daszkowski, T., Clark, E., Mcduff, P., Heffernan, C., Duffy, L., Poole, S., Ryll, T., Sheehy, P., Strachan, D., Beattie, D., Souquet, J., Pollard, D., Stauch, O., Bezy, P., Sauer, T., Boettcher, L., Simpson, C., Dakin, J., Pitt, S. & Boyle, A. (2017b), Biomanufacturing Technology Roadmap: Executive Summary, Technical report, BioPhorum Operations Group Ltd.

Scarff, M., Arnold, S. A., Harvey, L. M. & McNeil, B. (2006), 'Near infrared spectroscopy for bioprocess monitoring and control: Current status and future trends', *Critical Reviews in Biotechnology* **26**(1), 17–39.

Seth, G., Hamilton, R. W., Stapp, T. R., Zheng, L., Meier, A., Petty, K., Leung, S. & Chary, S. (2013), 'Development of a new bioprocess scheme using frozen seed train intermediates to initiate CHO cell culture manufacturing campaigns', *Biotechnology and Bioengineering* **110**(5), 1376–1385.

Shah, Y. T. (1979), *Gas Liquid Solid Reactor Design*, Vol. 327, McGraw-Hill International.

Sharma, C., Malhotra, D. & Rathore, A. S. (2011), 'Review of computational fluid dynamics applications in biotechnology processes', *Biotechnology Progress* **27**(6), 1497–1510.

Sherr, C. J. & Roberts, J. M. (1999), 'CDK inhibitors: Positive and negative regulators of G1-phase progression', *Genes & Development* **13**(12), 1501–1512.

Shukla, A. A. & Gottschalk, U. (2013), 'Single-use disposable technologies for biopharmaceutical manufacturing', *Trends in Biotechnology* **31**(3), 147–154.

Shukla, A. A. & Thömmes, J. (2010), 'Recent advances in large-scale production of monoclonal antibodies and related proteins', *Trends in Biotechnology* **28**(5), 253–261.

Sidoli, F. R., Asprey, S. P. & Mantalaris, A. (2006), 'A coupled single cell-population-balance model for mammalian cell cultures', *Industrial and Engineering Chemistry Research* **45**(16), 5801–5811.

References

Sidoli, F. R., Mantalaris, A. & Asprey, S. P. (2004), 'Modelling of mammalian cells and cell culture processes', *Cytotechnology* **44**(1–2), 27–46.

Sieblist, C., Jenzsch, M. & Pohlscheidt, M. (2013), 'Influence of pluronic F68 on oxygen mass transfer', *Biotechnology Progress* **29**(5), 1278–1288.

Sieblist, C., Jenzsch, M. & Pohlscheidt, M. (2016), 'Equipment characterization to mitigate risks during transfers of cell culture manufacturing processes', *Cytotechnology* **68**(4), 1381–1401.

Sieblist, C., Jenzsch, M., Pohlscheidt, M. & Lübbert, A. (2011), 'Insights into large-scale cell-culture reactors: I. Liquid mixing and oxygen supply', *Biotechnology Journal* **6**(12), 1532–1546.

Sieck, J. B., Budach, W. E., Suemeghy, Z., Leist, C., Villiger, T. K., Morbidelli, M. & Soos, M. (2014), 'Adaptation for survival: Phenotype and transcriptome response of CHO cells to elevated stress induced by agitation and sparging', *Journal of Biotechnology* **189**, 94–103.

Sieck, J. B., Cordes, T., Budach, W. E., Rhiel, M. H., Suemeghy, Z., Leist, C., Villiger, T. K., Morbidelli, M. & Soos, M. (2013), 'Development of a scale-down model of hydrodynamic stress to study the performance of an industrial CHO cell line under simulated production scale bioreactor conditions', *Journal of Biotechnology* **164**(1), 41–49.

Siganporia, C. C., Ghosh, S., Daszkowski, T., Papageorgiou, L. G. & Farid, S. S. (2014), 'Capacity planning for batch and perfusion bioprocesses across multiple biopharmaceutical facilities', *Biotechnology Progress* **30**, 594–606.

Singer, M. & Soll, D. (1973), 'Guidelines for DNA hybrid molecules', *Science* **181**, 1114.

Smelko, P. J., Wiltberger, R. K., Hickman, F. E., et al. (2011), 'Performance of high intensity fed-batch mammalian cell cultures in disposable bioreactor systems', *Biotechnology Progress* **27**(5), 1358–1364.

Sokolov, M., Morbidelli, M., Butté, A., Souquet, J. & Broly, H. (2018), 'Sequential multivariate cell culture modeling at multiple scales supports systematic shaping of a monoclonal antibody toward a quality target', *Biotechnology Journal* **13**(4), 1700461.

Sokolov, M., Ritscher, J., MacKinnon, N., Bielser, J.-M., Brühlmann, D., Rothenhäusler, D., Thanei, G., Soos, M., Stettler, M., Souquet, J., Broly, H., Morbidelli, M. & Butté, A. (2017a), 'Robust factor selection in early cell culture process development for the production of a biosimilar monoclonal antibody', *Biotechnology Progress* **33**(1), 181–191.

Sokolov, M., Ritscher, J., MacKinnon, N., Bielser, J.-M. J.-M., Brühlmann, D., Rothenhäusler, D., Thanei, G., Soos, M., Stettler, M., Souquet, J., Broly, H., Morbidelli, M. & Butté, A. (2017b), 'Robust factor selection in early cell culture process development for the production of a biosimilar monoclonal antibody', *Biotech Progress Journal* **33**(1), 181.191.

Sokolov, M., Ritscher, J., Mackinnon, N., Souquet, J., Broly, H., Morbidelli, M. & Butté, A. (2017c), 'Enhanced process understanding and multivariate prediction of the relationship between cell culture process and monoclonal antibody quality', *Biotechnology Progress* pp. 1–13.

Sokolov, M., Soos, M., Neunstoecklin, B., Morbidelli, M., Butté, A., Leardi, R., Solacroup, T., Stettler, M. & Broly, H. (2015), 'Fingerprint detection and process prediction by multivariate analysis of fed-batch monoclonal antibody cell culture data', *Biotechnology Progress* **31**(6), 1633–1644.

Soleas, G. J., Diamandis, E. P. & Goldberg, D. M. (1997), 'Wine as a biological fluid: History, production, and role in disease prevention', *Journal of Clinical Laboratory Analysis* **11**(5), 287–313.

Solomon, B. L. & Garrido-Laguna, I. (2018), 'TIGIT: A novel immunotherapy target moving from bench to bedside', *Cancer Immunology, Immunotherapy* **67**(11), 1659–1667.

References

Sommeregger, W., Sissolak, B., Kandra, K., von Stosch, M., Mayer, M. & Striedner, G. (2017), 'Quality by control: Towards model predictive control of mammalian cell culture bioprocesses', *Biotechnology Journal* **12**(7), 1–7.

Soos, M., Ehrl, L., Baĩ̀bler, M. U. & Morbidelli, M. (2010), 'Aggregate breakup in a contracting nozzle', *Langmuir* **26**(1), 10–18.

Soos, M., Kaufmann, R., Winteler, R., Kroupa, M. & Lüthi, B. (2013), 'Determination of maximum turbulent energy dissipation rate generated by a rushton impeller through large eddy simulation', *AIChE Journal* **59**(10), 3642–3658.

Sou, S. N., Jedrzejewski, P. M., Lee, K., Sellick, C., Polizzi, K. M. & Kontoravdi, C. (2017), 'Model-based investigation of intracellular processes determining antibody Fc-glycosylation under mild hypothermia', *Biotechnology and Bioengineering* **114**(7), 1570–1582.

Stahmann, K. P., Revuelta, J. L. & Seulberger, H. (2000), 'Three biotechnical processes using Ashbya gossypii, Candida famata, or Bacillus subtilis compete with chemical riboflavin production', *Applied Microbiology and Biotechnology* **53**(5), 509–516.

Steeno, G. S. (2010), 'Experimental design for pharmaceutical development', in J. David, ed., *Chemical Engineering in the Pharmaceutical Industry: R&D to Manufacturing*, John Wiley and Sons, pp. 597–620.

Steinebach, F., Angarita, M., Karst, D. J., Müller-Späth, T. & Morbidelli, M. (2016a), 'Model based adaptive control of a continuous capture process for monoclonal antibodies production', *Journal of Chromatography A* **1444**, 50–56.

Steinebach, F., Müller-Späth, T. & Morbidelli, M. (2016b), 'Continuous counter-current chromatography for capture and polishing steps in biopharmaceutical production', *Biotechnology Journal* **11**(9), 1126–1141.

Steinebach, F., Ulmer, N., Wolf, M., Decker, L., Schneider, V., Wälchli, R., Karst, D., Souquet, J. & Morbidelli, M. (2017), 'Design and operation of a continuous integrated monoclonal antibody production process', *Biotechnology Progress* **33**(5), 1303–1313.

Swann, P., Brophy, L., Strachan, D., Lilly, E. & Jeffers, P. (2017), Biomanufacturing Technology Roadmap: In-line monitoring and real-time release, Technical report, BioPhorum Operations Group Ltd.

Tabas, I. & Kornfeld, S. (1979), 'Purification and characterization of a rat liver Golgi alpha-mannosidase capable of processing asparagine-linked oligosaccharides', *The Journal of Biological Chemistry* **254**(22), 11655–63.

Takesono, S., Onodera, M., Toda, K., Yoshida, M., Yamagiwa, K. & Ohkawa, A. (2006), 'Improvement of foam breaking and oxygen-transfer performance in a stirred-tank fermenter', *Bioprocess and Biosystems Engineering* **28**(4), 235–242.

Tanzeglock, T., Soos, M., Stephanopoulos, G. & Morbidelli, M. (2009), 'Induction of mammalian cell death by simple shear and extensional flows', *Biotechnology and Bioengineering* **104**(2), 360–370.

Tao, Y., Shih, J., Sinacore, M., Ryll, T. & Yusuf-Makagiansar, H. (2011), 'Development and implementation of a perfusion-based high cell density cell banking process', *Biotechnology Progress* **27**(3), 824–829.

Teixeira, A., Cunha, A., Clemente, J., Moreira, J., Cruz, H., Alves, P., Carrondo, M. & Oliveira, R. (2005), 'Modelling and optimization of a recombinant BHK-21 cultivation process using hybrid grey-box systems', *Journal of Biotechnology* **118**(3), 290–303.

Thomas, T. N. (2017), 'Are we prepared to meet the demands of a challenging, but promising future?', in *Integrated Continuous Biomanufacturing III*, Suzanne Farid, University College London, United Kingdom Chetan Goudar, Amgen, USA Paula Alves, IBET, Portugal Veena Warikoo, Axcella Health, Inc., USA Eds, ECI Symposium Series'.

Tribe, L. A., Briens, C. L. & Margaritis, A. (1995), 'Determination of the volumetric mass transfer coefficient (kla) using the dynamic "gas out–gas in" method', *Biotechnology and Bioengineering* **46**, 388–392.

Tsang, V. L., Wang, A. X., Yusuf-Makagiansar, H. & Ryll, T. (2014), 'Development of a scale down cell culture model using multivariate analysis as a qualification tool', *Biotechnology Progress* **30**(1), 152–160.

Tziampazis, E. & Sambanis, A. (1994), 'Modeling of cell culture processes', *Cytotechnology* **14**(3), 191–204.

Umaña, P. & Bailey, J. E. (1997), 'A mathematical model of N-linked glycoform biosynthesis', *Biotechnology and Bioengineering* **55**(6), 890–908.

Undey, C., Low, D., Menezes, J. C. & Koch, M. (2011), *PAT Applied in Biopharmaceutical Process Development and Manufacturing: An Enabling Tool for Quality-by-Design*, Vol. 33, CRC Press.

Van't Riet, K. (1979), 'Review of measuring methods and results in nonviscous gas–liquid mass transfer in stirred vessels', *Industrial and Engineering Chemistry Process Design and Development* **18**(3), 357–364.

Velasco, A. (1993), 'Cell type-dependent variations in the subcellular distribution of alpha-mannosidase I and II', *The Journal of Cell Biology* **122**(1), 39–51.

Versteeg, H. K. & Malalasekera, W. (1995), *An Introduction to Computational Fluid Dynamics*, John Wiley and Sons.

Villiger, T. K., Morbidelli, M. & Soos, M. (2015), 'Experimental determination of maximum effective hydrodynamic stress in multiphase flow using shear sensitive aggregates', *AIChE Journal* **61**(5), 1735–1744.

Villiger, T. K., Neunstoecklin, B., Karst, D. J., Lucas, E., Stettler, M., Broly, H., Morbidelli, M. & Soos, M. (2018), 'Experimental and CFD physical characterization of animal cell bioreactors: From micro- to production scale', *Biochemical Engineering Journal* **131**, 84–94.

Villiger, T. K., Roulet, A., Périlleux, A., Stettler, M., Broly, H., Morbidelli, M., Soos, M., Scibona, E., Stettler, M., Broly, H., Morbidelli, M., Soos, M., Roulet, A., Périlleux, A., Stettler, M., Broly, H., Morbidelli, M. & Soos, M. (2016a), 'Controlling the time evolution of mAb N-linked glycosylation – Part I: Micro-bioreactor experiments', *Biotechnology Progress* **32**(5), 1123–1134.

Villiger, T. K., Scibona, E., Stettler, M., Broly, H., Morbidelli, M. & Soos, M. (2016b), 'Controlling the time evolution of mAb N-linked glycosylation – Part II: Model-based predictions', *Biotechnology Progress* **32**(5), 1135–1148.

Villiger, T. K., Steinhoff, R. F., Ivarsson, M., Solacroup, T., Stettler, M., Broly, H., Krismer, J., Pabst, M., Zenobi, R., Morbidelli, M. & Soos, M. (2016c), 'High-throughput profiling of nucleotides and nucleotide sugars to evaluate their impact on antibody N-glycosylation', *Journal of Biotechnology* **229**, 3–12.

Villiger-Oberbek, A., Yang, Y., Zhou, W. & Yang, J. (2015), 'Development and application of a high-throughput platform for perfusion-based cell culture processes', *Journal of Biotechnology* **212**, 21–29.

Vogg, S., Müller-Späth, T. & Morbidelli, M. (2018), 'Current status and future challenges in continuous biochromatography', *Current Opinion in Chemical Engineering* **22**, 138–144.

Voisard, D., Meuwly, F., Ruffieux, P. A., Baer, G. & Kadouri, A. (2003), 'Potential of cell retention techniques for large-scale high-density perfusion culture of suspended mammalian cells', *Biotechnology and Bioengineering* **82**(7), 751–765.

Vojinović, V., Cabral, J. M. S. & Fonseca, L. P. (2006), 'Real-time bioprocess monitoring: Part I: In situ sensors', *Sensors and Actuators, B: Chemical* **114**, 1083–1091.

References

Von Stosch, M., Hamelink, J.-M. & Oliveira, R. (2016), 'Hybrid modeling as a QbD/PAT tool in process development: An industrial E. coli case study', *Bioprocess and Biosystems Engineering* **39**(5), 773–784.

Von Stosch, M. & Willis, M. J. (2017), 'Intensified design of experiments for upstream bioreactors', *Engineering in Life Sciences* **17**(11), 1173–1184.

Vulto, A. G. & Jaquez, O. A. (2017), 'The process defines the product: What really matters in biosimilar design and production?', *Rheumatology* **56**(suppl 4), 14–29.

Walsh, G. (2010), 'Post-translational modifications of protein biopharmaceuticals', *Drug Discovery Today* **15**(17–18), 773–780.

Walsh, G. (2014), 'Biopharmaceutical benchmarks 2014', *Nature Biotechnology* **32**(7), 992–1000.

Walther, J., Godawat, R., Hwang, C., Abe, Y., Sinclair, A. & Konstantinov, K. (2015), 'The business impact of an integrated continuous biomanufacturing platform for recombinant protein production', *Journal of Biotechnology* **213**, 3–12.

Walther, J., Lu, J., Hollenbach, M., Yu, M., Hwang, C., McLarty, J. & Brower, K. (2018), 'Perfusion cell culture decreases process and product heterogeneity in a head-to-head comparison with fed-batch', *Biotechnology Journal* **14**(2), 1700733.

Walther, J., McLarty, J. & Johnson, T. (2019), 'The effects of alternating tangential flow (ATF) residence time, hydrodynamic stress, and filtration flux on high-density perfusion cell culture', *Biotechnology and Bioengineering* **116**(2), 320–332.

Walther, J., Shah, N., Hollenbach, M., Wang, J., Yu, M., Lu, J., Yang, Y., Konstantinov, K. B. & Hwang, C. (2016), 'Overcoming process intensification challenges to deliver a manufacturable and competitive integrated continuous biomanufacturing platform', in *Cell Culture Engineering XV*, Robert Kiss, Genentech Sarah Harcum, Clemson University Jeff Chalmers, Ohio State University Eds, ECI Symposium Series.

Wang, J., Liu, L., Ball, T., Yu, L., Li, Y. & Xing, F. (2016), 'Revealing a 5,000-y-old beer recipe in China', *Proceedings of the National Academy of Sciences* **113**(23), 6444–6448.

Wang, S. B., Lee-Goldman, A., Ravikrishnan, J., Zheng, L. & Lin, H. (2018), 'Manipulation of the sodium-potassium ratio as a lever for controlling cell growth and improving cell specific productivity in perfusion CHO cell cultures', *Biotechnology and Bioengineering* **115**(4), 921–931.

Wang, S., Godfrey, S., Ravikrishnan, J., Lin, H., Vogel, J. & Coffman, J. (2017), 'Shear contributions to cell culture performance and product recovery in ATF and TFF perfusion systems', *Journal of Biotechnology* **246**, 52–60.

Wang, Z., Zhuge, J., Fang, H. & Prior, B. A. (2001), 'Glycerol production by microbial fermentation: A review', *Biotechnology Advances* **19**(3), 201–223.

Warikoo, V., Godawat, R., Brower, K., Jain, S., Cummings, D., Simons, E., Johnson, T., Walther, J., Yu, M., Wright, B., McLarty, J., Karey, K. P., Hwang, C., Zhou, W., Riske, F. & Konstantinov, K. (2012), 'Integrated continuous production of recombinant therapeutic proteins', *Biotechnology and Bioengineering* **109**(12), 3018–3029.

Watanabe, S., Shuttleworth, J. & Al-Rubeai, M. (2002), 'Regulation of cell cycle and productivity in NS0 cells by the over-expression of p21CIP1', *Biotechnology and Bioengineering* **77**(1), 1–7.

Webster, T. A., Hadley, B. C., Hilliard, W., Jaques, C. & Mason, C. (2018), 'Development of generic raman models for a GS-KO^{TM} CHO platform process', *Biotechnology Progress* **34**(3), 730–737.

Whelan, J., Craven, S. & Glennon, B. (2012), 'In situ Raman spectroscopy for simultaneous monitoring of multiple process parameters in mammalian cell culture bioreactors', *Biotechnology Progress* (5), 1355–1362.

References

Whitford, W. G. (2014), 'Single-use systems support continuous bioprocessing by perfusion culture', in *Continuous Processing in Pharmaceutical Manufacturing*, Wiley-VCH Verlag GmbH and Co. KGaA, Weinheim, chapter 9, pp. 183–226.

Wold, S., Sjöström, M. & Eriksson, L. (2001), 'PLS-regression: A basic tool of chemometrics', *Chemometrics and Intelligent Laboratory Systems* **58**(2), 109–130.

Wolf, M. K. F., Closet, A., Bzowska, M., Bielser, J.-M., Souquet, J., Broly, H. & Morbidelli, M. (2019a), 'Improved performance in mammalian cell perfusion cultures by growth inhibition', *Biotechnology Journal* **14**(2), 1700722.

Wolf, M. K. F., Lorenz, V., Karst, D. J., Souquet, J., Broly, H. & Morbidelli, M. (2018), 'Development of a shake tube-based scale-down model for perfusion cultures', *Biotechnology and Bioengineering* **115**(11), 2703–2713.

Wolf, M. K. F., Müller, A., Souquet, J., Broly, H. & Morbidelli, M. (2019b), 'Process design and development of a mammalian cell perfusion culture in shake-tube and benchtop bioreactors', *Biotechnology and Bioengineering* **116**(8), 1973–1985.

Wolf, M. K. F., Pechlaner, A., Lorenz, V., Karst, D. J., Souquet, J., Broly, H., & Morbidelli, M. (2019c). A two-step procedure for the design of perfusion bioreactors. *Biochemical Engineering Journal*, **151**, 107295.

Wolton, A. D. & Rayner, A. (2014), 'Lessons learned in the ballroom', *Pharmaceutical Engineering* **34**(4), 1–5.

Wong, Y. H., Krishnaswamy, P. R. & Teo, W. K. (1992), 'Advanced control of pH in mammalian cell culture', in S. Furusaki, I. Endo & R. Matsuno, eds, *Biochemical Engineering for 2001*, Springer Japan, pp. 689–691.

Woodcock, J. (2014), 'Modernizing pharmaceutical manufacturing: Continuous manufacturing as a key enabler, in *International Symposium on Continuous Manufacturing of Pharmaceuticals, Cambridge, MA*.

Wright, B., Bruninghaus, M., Vrabel, M., Walther, J. & Shah, N. (2015), 'A novel seed-train process', *BioProcess International* **13**(3), 16–25.

Wurm, F. M. (2004), 'Production of recombinant protein therapeutics in cultivated mammalian cells', *Nature Biotechnology* (11), 1393–1398.

Xu, P., Clark, C., Ryder, T., Sparks, C., Zhou, J., Wang, M., Russell, R. & Scott, C. (2017a), 'Characterization of TAP Ambr 250 disposable bioreactors, as a reliable scale-down model for biologics process development', *Biotechnology Progress* **33**(2), 478–489.

Xu, S. & Chen, H. (2016), 'High-density mammalian cell cultures in stirred-tank bioreactor without external pH control', *Journal of Biotechnology* **231**, 149–159.

Xu, S., Gavin, J., Jiang, R. & Chen, H. (2017b), 'Bioreactor productivity and media cost comparison for different intensified cell culture processes', *Biotechnology Progress* **33**(4), 867–878.

Xu, S., Jiang, R., Chen, Y., Wang, F. & Chen, H. (2017c), 'Impact of Pluronic® F68 on hollow fiber filter-based perfusion culture performance', *Bioprocess and Biosystems Engineering* **40**(9), 1317–1326.

Yang, W. C., Lu, J., Kwiatkowski, C., Yuan, H., Kshirsagar, R., Ryll, T. & Huang, Y. M. (2014), 'Perfusion seed cultures improve biopharmaceutical fed-batch production capacity and product quality', *Biotechnology Progress* **30**(3), 616–625.

Yang, W. C., Minkler, D. F., Kshirsagar, R., Ryll, T. & Huang, Y. M. (2016), 'Concentrated fed-batch cell culture increases manufacturing capacity without additional volumetric capacity', *Journal of Biotechnology* **217**, 1–11.

Yao, T. & Asayama, Y. (2017), 'Animal-cell culture media: History, characteristics, and current issues', *Reproductive Medicine and Biology* **16**(2), 99–117.

References

Yeung, K. S. Y., Hoare, M., Thornhill, N. F., Williams, T. & Vaghjiani, J. D. (2002), 'NearâŁinfrared spectroscopy for bioprocess monitoring and control', *Biotechnology and Bioengineering* **63**(6), 684–693.

Yoon, S. K., Choi, S. L., Song, J. Y. & Lee, G. M. (2005), 'Effect of culture pH on erythropoietin production by Chinese hamster ovary cells grown in suspension at 32.5 and 37.0C', *Biotechnology and Bioengineering* **89**(3), 345–356.

Yoon, S. & Konstantinov, K. B. (1994), 'Continuous, real-time monitoring of the oxygen uptake rate (OUR) in animal cell bioreactors', *Biotechnology and Bioengineering* **44**, 983–990.

Yu, L. X., Baker, J., Berlam, S. C., Boam, A., Brandreth, E. J., Buhse, L., Cosgrove, T., Doleski, D., Ensor, L., Famulare, J., Ganapathy, M., Grampp, G., Hussong, D., Iser, R., Johnston, G., Kesisoglou, F., Khan, M., Kozlowski, S., Lacana, E., Lee, S. L., Miller, S., Miksinski, S. P., Moore, C. M. V., Mullin, T., Raju, G. K., Raw, A., Rosencrance, S., Rosolowsky, M., Stinavage, P., Thomas, H., Wesdyk, R., Windisch, J. & Vaithiyalingam, S. (2015), 'Advancing product quality: A summary of the inaugural FDA/PQRI Conference', *The AAPS Journal* **17**(4), 1011–1018.

Zagari, F., Jordan, M., Stettler, M., Broly, H. & Wurm, F. M. (2013), 'Lactate metabolism shift in CHO cell culture: The role of mitochondrial oxidative activity', *New Biotechnology* **30**(2), 238–245.

Zalai, D., Tobak, T. & Putics, Á. (2015), 'Impact of apoptosis on the on-line measured dielectric properties of CHO cells', *Bioprocess and Biosystems Engineering* **38**(12), 2427–2437.

Zhang, A., Tsang, V. L., Moore, B., Shen, V., Huang, Y. M., Kshirsagar, R. & Ryll, T. (2015a), 'Advanced process monitoring and feedback control to enhance cell culture process production and robustness', *Biotechnology and Bioengineering* **112**(12), 2495–2504.

Zhang, Y. H. P., Sun, J. & Ma, Y. (2016), 'Biomanufacturing: History and perspective', *Journal of Industrial Microbiology and Biotechnology* **44**(4–5), 773–784.

Zhang, Y., Stobbe, P., Silvander, C. O. & Chotteau, V. (2015b), 'Very high cell density perfusion of CHO cells anchored in a non-woven matrix-based bioreactor', *Journal of Biotechnology* **213**, 28–41.

Zhou, W. & Kantardjieff, A. (2014), *Mammalian Cell Cultures for Biologics Manufacturing*, Springer.

Zhu, L. K., Song, B. Y., Wang, Z. L., Monteil, D. T., Shen, X., Hacker, D. L., De Jesus, M. & Wurm, F. M. (2017), 'Studies on fluid dynamics of the flow field and gas transfer in orbitally shaken tubes', *Biotechnology Progress* **33**(1), 192–200.

Zhu, M. M., Goyal, A., Rank, D. L., Gupta, S. K., Vanden Boom, T. & Lee, S. S. (2005), 'Effects of elevated pCO_2 and osmolality on growth of CHO cells and production of antibody-fusion protein B1: A case study', *Biotechnology Progress* **21**(1), 70–77.

Zoro, B. & Tait, A. (2017), 'Development of a novel automated perfusion mini bioreactor ambr® 250 perfusion', in *Integrated Continuous Biomanufacturing III*, Suzanne Farid, University College London, United Kingdom Chetan Goudar, Amgen, USA Paula Alves, IBET, Portugal Veena Warikoo, Axcella Health, Inc., USA Eds, ECI Symposium Series 2017', p. 250.

Index

Index

Index

Index

Printed in the United States
By Bookmasters